Train Your Dog Step-By-Step

3 BOOKS IN 1

Learn How To Train Your Dog, Tips And Tricks, Techniques And Strategies For The Best Dog Ever

PAUL DAVIS

© **Copyright 2019 - All rights reserved.**

The content contained within this book may not be reproduced, duplicated, or transmitted without direct written permission from the author or the publisher.

Under no circumstances will any blame or legal responsibility be held against the publisher or author for any damages, reparation, or monetary loss due to the information contained within this book, either directly or indirectly.

Legal Notice:

This book is copyright protected. It is only for personal use. You cannot amend, distribute, sell, use, quote, or paraphrase any part or the content within this book, without the consent of the author or publisher.

Disclaimer Notice:

Please note the information contained within this document is for educational and entertainment purposes only. All effort has been executed to present accurate, up to date, reliable, complete information. No warranties of any kind are declared or implied. Readers acknowledge that the author is not engaging in the rendering of legal, financial, medical, or professional advice. The content within this book has been derived from various sources. Please consult a licensed professional before attempting any techniques outlined in this book.

By reading this document, the reader agrees that under no circumstances is the author responsible for any losses, direct or indirect, that are incurred as a result of the use of information contained within this document, including, but not limited to, errors, omissions, or inaccuracies.

Table of Contents

BOOK 1: E-COLLAR TRAINING STEP-BY-STEP 11

Introduction 13

Chapter 1: Choosing the Right Dog for You 16

   Tips for Choosing the Best Dog for You 16

      Choosing the Best Breed for Training 19

      Best Breeds for Training 20

      Where to Go? 25

      Tips to Keep in Mind Before Bringing Your Dog Home ... 28

Chapter 2: E-Collar Basics 31

      What Is the E-Collar? 31

      Types of E-Collars 32

      E-Collar Accessories 35

      How to Use an E-Collar 35

      E-Collar Safety 39

      Benefits of an E-Collar 40

      Common Myths About the E-Collar 42

Chapter 3: Your Dog and Their E-Collar 44

      Choosing the Best E-Collar 44

      The Fundamental Five 44

      Your Dog's Reaction to the E-Collar 46

Chapter 4: What You Need to Know Before Training Begins ... 53

      General Training Tips for Dogs at Any Age 53

      Training Your Puppy 63

      Training Your Older Dog 66

Chapter 5: Let the Training Begin 69
Tips to Prepare Your Dog for Training 69
Basic Commands 70
Home Base and Perimeter Training 72
Sit Training..76
Lay Down Training..78
Come Training 79
Stay Training 81
Get Down Training 82

Chapter 6: Training Strategies, Levels, and Your Dog 84
Training Levels 84
Practice Real Life Training 87
Intermediate Level Tricks for Dogs 94
Spin and Twist.. 98

Chapter 7: Advanced Training with E-Collars 101
Will I Stop Training?... 101
Readjusting the E-Collar for a Growing Dog.......................................102
Agility Training..102
Basic Training Before the E-Collar 103
Boing 106
Hunting...107

Chapter 8: Common Mistakes 109
Lack of Consistency in Training..109
Training Your Dog for Too Long...109
Dog Owners Don't Send a Signal Immediately 110
People Become Codependent on the E-Collar......................................111
The Dog Has Not Received Any Type of Prior Training.....................111
People Don't Understand How the E-Collar Works 112

Chapter 9: Frequently Asked Questions and Answers 114
    Question #1: Do Different Breeds of Dogs Learn Differently? 114
    Questions #2: Do I Wean My Dog Off the E-Collar? 114
    Question #3: How Do I Know the E-Collar Isn't Harming My Dog? 115
    Question #4: How Do I Know When to Start Using the E-collar? 115
    Question #5: I Have Seen Other Dogs React Negatively to the Shock from the E-Collar. How Do I Know My Dog Won't? 116

Conclusion 117

References 121

BOOK 2: TIPS AND TRICKS TO DOG TRAINING 127

Introduction 129

Chapter 1: The Basics 131
    What to Know Before Training 131
    Common Dog Training Methods 137

Chapter 2: Training and Your Dog 144
    Dog Breeds and Training 144
    Caring for Your Dog 148
    Tips to Overcome Training Challenges 151
    Different Ways to Reward Your Dog 154
    Preparing Your Dog for Training 155
    Keys for Successful Dog Training 157

Chapter 3: Tips and Tricks for Training Your Puppy 159
    Know Your Puppy Before Training Begins 159
    Tips for Training Your Puppy 161
    Understand Your Puppy's Developmental Stages 163
    Foundation Training 166

Chapter 4: Tips and Tricks for Training Your Adolescent Dog 173

What You Need to Know About Your Adolescent Dog 173

Basic Training.. 176

Real Life Training 182

Beyond Basic Training 183

Chapter 5: Tips and Tricks for Training Your Adult Dog................... 194

Keeping Your Adult Dog Healthy 194

Adult Dog Training Techniques... 196

Training Tips for Older Dogs 197

Chapter 6: Tips and Tricks for Training Your Senior Dog 207

Keeping a Senior Dog Healthy 207

Training Techniques for Senior Dogs.. 209

Chapter 7: So, You Want to Train Your Dog to Dance and Other Tricks...... 214

Keeping Your Dog Active 214

Always Remember Safety 214

Traditional Favorites 215

Silly Dog Tricks 218

Expert Tricks 219

Dancing 221

Chapter 8: Common Mistakes 223

Conclusion 228

References.. 232

BOOK 4 TRAINING YOUR PUPPY STEP-BY-STEP 239

Introduction 241

Chapter 1: Choosing the Right Dog... 243

Picking Where Your Dog Comes From.. 244

Energy Levels of Certain Dogs 247

Child-Friendly or Not 252

Interacting with Other Animals 254

Grooming Needs 259

Other Breed Specific Conditions to Remember 261

Chapter 2: Preparing Your Home for Their Arrival 264

Starting Positive Habits from the Beginning 264

The Right House Setting 267

Tools Needed 269

The Training Room for Your Dog 271

Hiding Certain Belongings 276

Choosing the Right Toys 278

Chapter 3: Principles to Train Dogs 281

Potty Training 281

Basic Training Rules 286

Positive Reinforcement and Remaining Patient 288

Specific Rules for Small Dogs 292

Specific Rules for Medium Dogs 293

Specific Rules for Large Dogs 294

Chapter 4: Training Your Puppy Outside of Your Home 295

Going for Walks and Leash Training 295

Doggy Daycare 299

Parks 301

Friends' Homes and New Environments 304

What to Do When You are Not Home 306

Public Tips and Reminders 308

Chapter 5: Skills that Dogs Need to Know 310

Sitting and Laying Down 310

Bark Control 313

Greeting Visitors and Walking Properly 315

    Knowing Their Name and Coming on Command 317

    Fun Tricks for Your Dog 319

Chapter 6: Kinds of Exercises for Your Puppy 322

    Training and Agility Classes 322

    Swimming 324

    Catch 326

    Dog-Led Walks 328

    Creating a Routine 328

Chapter 7: What to Do for the Health of Your Dog 331

    Spay and Neutering 334

    Picking the Right Food and Treats 336

    Avoiding People Food 338

    Flea Treatments and Bathing 340

Chapter 8: What Not to Do When Training Dogs 344

    Aggression in Training 344

    Inconsistencies 347

    Useless Repetition 348

    Confusing the Dog 349

    Abandoning a Ritual or Habit 351

    Treating All Dogs, the Same 352

    Positive Reinforcement with Bad Behavior 353

    Not Proofing Tricks 354

    Waiting Too Long 356

Chapter 9: Your Step-by-Step Training Plan 357

    Get to Know Your Dog 357

    Create a Schedule 357

    Keep it Simple 359

    Work on Rewards 360

    Repeat, Repeat, Repeat.. 360

    Check-In with Your Dog 360

Conclusion 362

References..365

BOOK 1

E-COLLAR TRAINING STEP-BY-STEP

A How-To Innovative Guide to Positively Train Your Dog through E- collars; Tips and Tricks and Effective Techniques for Different Species of Dogs

PAUL DAVIS

Introduction

You are at your local animal shelter looking through the cages. You walk by slowly, talking to each dog. "Hey there, how are you doing today?" you say to each one of the dogs as you read the little notecard of information attached to the cage. It's so hard to choose one dog to bring home. You want to bring them all home. But you know that you need to pick the best dog for yourself, your family, and think of the dog.

Living in an apartment, you realize a smaller dog is best. It will get exercise by running around the apartment easier than a larger dog will. You also know you need a dog that is comfortable around kids, won't get lonely easily, and is easily trainable. With you and your husband working full-time jobs, you can't be home all the time for your new pup.

Through your research, you know many breeds of smaller dogs are easily trainable. Walking by a cage, you see a Miniature Schnauzer named Phillip. You smile at Phillip as you read information about him. He is three years old, neutered, enjoys children, and has received basic training. Through your conversation with the shelter employee, you find out that the employees are training Phillip with an e-collar or electronic collar. You know very little about this device, so the employee shows you how it works.

This book is going to take you through everything you need to know about e-collar training and your dog. You will not only learn the basics of e-collar training, but you'll learn how to help your dog thrive. Diving into Chapter 1, you will get basic information, such as easily trainable dog breeds, what factors to think of when you are choosing the best dog for you, how to care for your dog, and where to go to choose your dog.

Chapter 2 will give you information on e-collar basics. After learning what an e-collar is, you can discover the types of e-collars and accessories available for your dog. This chapter will also give you a step-by-step guide on how to use an e-collar. Of course, you can't learn how to use an e-collar without learning about the safety of e-collars. If you are still unsure if e-collar is the way to go, Chapter 2 debunks several myths that people have about e-collars.

Deciding that you want the e-collar, you wonder how it will affect your dog. Is there a certain e-collar you should choose? What will your dog's reaction be? Instead of wondering about these questions, it's time to read Chapter 3. In this chapter, you will learn about the fundamental five when it comes to picking the right e-collar for your dog. You will also read about a couple of case examples when it comes to the unintended consequences of an e-collar your dog can have if you do not train them properly.

Chapter 4 gives you everything you need to know before you begin training. You will receive general tips on training a dog at any age, information on obedience classes, and how your dog's age matters when it comes to training. To give you the greatest amount of information possible so you are prepared to start training your dog, this chapter gives you tips for training your puppy and tips for training an older dog.

Finally, in Chapter 5, you can let the training begin. This chapter starts with giving you tips to prepare your dog for training. You will then get a step by step guide on how to train your dog with the basic commands, such as bed training, home base training, sitting, laying down, and getting down.

Chapter 6 goes a little further into training by looking at the training levels you will reach if you enroll your dog in obedience school. Of course, you can always focus on training your dog through these levels in your home. One of the best parts of this chapter is that it gives you some real life training examples so you know what to do when you are practicing training your dog in a real setting, such as bringing them on a car ride or to a dog-friendly store. You will also receive information on how to handle aggressiveness from dogs and some intermediate level tricks, such as teaching your dog to play fetch.

Book 1 - E-Collar Training Step-By-Step

The advanced training shows up in Chapter 7. Not only is the popular question of "Will I ever stop training my dog" be answered, but you will read tips about agility training and hunting. You can also learn a couple of fun tricks to see if your dog would be great at agility training.

What are the most common mistakes dog owners make and how you can pay attention so you don't make the same mistakes? You can find out all this information in Chapter 8. You will learn how discipline doesn't really help your dog, though most people incorporate time out when their dog does something like scratch up the wall.

This book will round off with Chapter 9, which discusses a number of common questions people have about dog training. Of course, you will get the answers to these questions!

Now that I have given you the rundown for what you are about to learn, it's time to dive into the wonderful information!

Chapter 1: Choosing the Right Dog for You

Choosing the right dog can be a big decision. While sometimes it seems that the right dog just "falls into your lap," other times you search for days or weeks. There are a lot of factors that influence your decision in getting a dog. For example, if you don't have a lot of money to put toward a dog, you look for a free dog or one from the shelter that is already fixed. If you live in an apartment, you may feel it is easier to get a small dog.

Tips for Choosing the Best Dog for You

With over 200 breeds of dogs, it's impossible to discuss all of them. Instead, I want to focus on tips that will help you choose the best dog for you.

Dogs Are Not Cheap

One factor to consider is how much money you can put toward a dog. While they are cheaper than children, they still need to go to the vet, they need food, toys, a crate, a leash, bowls for their food and water, grooming, and training tools or classes. Just because you got a dog for free from a social media site or your neighbor, does not mean that you will never put money toward your dog. Plus, most people like to spoil their dog with a sweater, especially if they live in colder temperatures.

Plan for Your Dog's Arrival

It is important to not just bring a dog home without any preparation. You want to prepare a place for your dog just like you would any member of your family. Just like humans, animals have feelings. They also need to feel a certain way to grow within their new home.

Book 1 - E-Collar Training Step-By-Step

If you have an older dog, you won't have to worry about "puppy proofing" your home. However, you will still want to ensure that you pick up anything the dog can harm themselves with. Even senior dogs will find interest in a screwdriver or want to taste the chocolate within their reach.

If you are bringing home a puppy, you will want to make sure that you puppy proof your entire home–you will be amazed at what a puppy can get into. For instance, you will want to ensure all your valuables are not in the puppy's reach. Remember, your puppy will be full of energy and into jumping. You will also want to secure any wires because puppies love to chew on anything.

You also need to think about what the dog needs. Some of the factors to think about are:

- *Security.* Just like people, one of the biggest ways to make a dog feel at home is to make them feel safe.
- *Love.* You will need to spend time with your dog and make sure they are adjusting well. They will need a lot of physical touch and attention. This will help calm their fears and know they are a welcomed member of the family.
- *Calming environment.* Dogs, especially puppies, become overwhelmed when there is a lot of chaos. They don't know what to do or how to handle it. This can make them act out, causing other problems.
- *Their space.* Just like people, dogs need their space. They need an area they can call their own. This is an area where they should feel safe and can have some dog time, just like humans need "me time."
- *They need supervision.* If you have a child and get a puppy, you will find a lot of common characteristics between the two, especially if you have a toddler and a puppy. While older dogs don't need as much supervision, a puppy is going to need a lot of attention and supervision. Puppies can become easily wound up and don't understand they can hurt someone if they jump on them. Therefore, watch your pup when they are playing to ensure everyone stays happy and safe.

You Need to Exercise Your Dog

If you are not one for walking, jogging or running, you will need to think about a dog walker or start exercising. Dogs need a lot of exercise, including walking every day. Dogs will need room so they can run, play, and jump. Therefore, you should always think twice about having a large dog in a small apartment where they rarely get to go outside.

Fortunately, if you are not one for exercise but still want a dog, there are breeds of dogs that require little exercise. For example, senior dogs will still need to go outside for a walk, but they are tamer and have lost a lot of energy. Smaller dogs are often okay to have if you are not huge on outdoor exercise because your home will give them room to play around and exercise.

Most Dogs Live for 10 Years or More

When you choose a dog, you want to keep in mind that you will most likely have this dog for over a decade. While this is usually a plus when it comes to picking a dog, some people are not interested in having a dog for years for various reasons. If this is the case, one of the best steps to take is to adopt an older dog. While you will need to factor in health problems, they might be the best choice if you only want a dog for a few years.

At the same time, you might want a dog who is known for a long lifespan. For example, you are getting a dog for your 10-year-old child to teach them responsibility. It takes years for most children to learn how to be responsible. At the same time, you will want a younger dog because you want your child and the dog to spend years together.

Most Dogs Require Grooming

The amount of grooming your dog needs will vary depending on the breed. Some dogs tend to need their fur trimmed often, while the fur on other breeds is fine without trimming. If you have a short-haired dog, you won't need to send them to the groomers as often. However, long-haired dogs require regular grooming and everyday brushing.

All dogs will need to get a bath from time to time. You need to wash their coat with special soap that is not only safe for the dogs, but also valuable for their fur. Along with fur comes shedding. Most dog breeds shed, but short-haired dogs are going to shed more than long-haired dogs.

A Puppy Doesn't Stay Small Forever

Puppies are adorable. Most people seem to want a puppy for several reasons. They are small, will lay in your lap, playful, and silly. Unfortunately, puppies do not stay small forever. Before you decide on a puppy because of its size, do a little research on the breed to see how big they can grow. Even puppies who seem on the smaller side for their breed can grow into large dogs. Don't let the adorable puppy size fool you!

Can You Meet the Parents?

Some people, especially people who want a show dog, will only take home a puppy if they can meet the puppy's parents. While this might not be important to you, it is something to consider if you have children, want to train your dog to become a show dog, or simply want to meet the parents. Parents can tell you a lot about how your puppy could grow up. But you always need to remember, a lot of your dog's temperament is going to deal with how you treat them.

Choosing the Best Breed for Training

It's true, some dogs are easily trainable. If you know that you will train your dog and don't want too many struggles with it, you will want to look at breeds that are easy to train.

What Makes a Dog Easily Trainable?

There are many factors that you need to think of when looking for an easily trainable dog.

- *Are they easily distracted?* Some dogs are more distractible than other dogs. For example, if you want a dog for hunting, you will not want to get a dog that is going to get distracted by leaves rustling in the trees or blowing in the wind. This can cause you to lose your target. You want a dog that is going to be aware of their surroundings yet stays focused on its target. Strong concentration also helps when you are training. Your dog will stay focused on what you are telling them and your hand signals.

- *What is the dog's personality?* Each breed has a distinct personality. Of course, each dog is a bit different than the others, but each breed has certain personality characteristics. For example, some breeds are going to cooperate better than other breeds.
- *What is the dog's instinct drives?* Each breed has certain instincts that drive them. This means these drives take over the dog's other instincts. For example, bloodhounds are ruled by their nose. Therefore, what they hear and see at the same time they are smelling something is not going to matter as much.

Best Breeds for Training

Miniature Schnauzer

As members of the Terrier group, Miniature Schnauzer's are not considered easily trainable dogs. This is because most Terrier dogs are harder to train than other breeds. However, Miniature Schnauzer's enjoy pleasing their owners and will work hard to accomplish a task. They are fearless, playful, and have a better trainability rate than any other dog in the Terrier family.

These small dogs can go anywhere you are. They can roam around on a farm in the country or remain content in an apartment. They have a spunky personality and make people laugh with their human-like expressions. Miniature Schnauzer gets along well with other dogs and children. They tend to make great watchdogs.

Doberman Pinscher

The Doberman Pinscher is a loyal and fearless companion. They are one of the breeds that are used for police and military training. They listen well, have great attention to detail, and the ability to concentrate on a task. Doberman Pinschers understand their commands, are quick to learn, and retain a lot of information.

Doberman Pinschers are not commonly chosen to be companions because they have a stereotype of being ruthless and known to attack. In reality, these dogs are cuddly, love to play, and friendly. They will only attack when they are threatened, like most dogs, or are trained and told to attack.

Doberman owners are honest about the stereotype their companions deal with on a daily basis. Walking your Doberman down the street, you start to notice people crossing the street ahead of you. They are watching how close you get as they cross. A few minutes later, you sit on a park bench to get a frisbee out of your bag. The guy next to you says, "I heard those dogs would attack their owner after their brain stops growing. Aren't you afraid of that dog attacking your neck?" You look at the person, get down on your knees, and say, "Come here, come give me a hug." You Doberman companion gently places both of their front paws on your shoulder, and you embrace them in a hug. Looking back at the person you say, "Not at all. They are big teddy bears."

Andrew Wildesen wrote an article dedicated to the stereotype Doberman Pinschers and their owners face. Wildesen stated that Dobermans are similar to every other dog–they can be trained to attack. However, they won't innately attack. Instead, they will bring more love and happiness into your life. Wildesen states, "The true Doberman is a lover with loads of enthusiasm… They are goofy and are guaranteed to make you laugh… they are tremendously loyal and love to cuddle… yes, Dobermans… are serious CUDDLE BUGS!" (Wildesen, n.d.).

One of the factors to be aware of with a Doberman is you need to put a lot of effort into their training. If you are looking for a dog that is easily trainable but doesn't require a lot of time, you will want to look at a different breed. You need to provide constant leadership and training for the Doberman breed to thrive.

Poodle

Poodles are common dogs chosen for many reasons. First, they are smaller in size, allowing them to adapt to various types of living situations. Poodles come in three different sizes: toy, miniature, and standard. Second, they are typically energetic and great with kids. Third, they are on every list that describes easily trainable dogs.

People love to train Poodles because they are extremely smart. However, some owners say they are too smart for their own good as they are known to outwit their trainers! They learn their tricks quickly and are eager to please their owner. It doesn't matter what size of Poodle you choose, they will all do their best to surprise you with the tricks you teach them.

Rottweiler

The Rottweiler breed is similar to Doberman Pinschers. They are known for their aggressive nature, but it is all in their training. On top of this, Rottweilers are easily trainable dogs because it is in their nature. In fact, they crave training and it is essential for their happiness.

Rottweilers are great to use as service and police dogs. However, they not only need basic training, they also need social training. This type of training goes beyond sitting and shaking hands as they like to be challenged.

German Shepherd

The German Shepherd is the Jack-of-all-trades dog. They are easy to train because they understand commands. They remember the tricks they are taught and need jobs to keep themselves happy. Furthermore, they are willing to learn and want to please you. They will go above and beyond the call of duty. German Shepherds tend to work tirelessly and are great for military and police work.

Like most dogs, German Shepherds need a lot of exercises. You will need to keep their mind going as well as their body. You don't need to wait too long to start training these dogs. They can understand the basic commands as young as eight weeks old.

Havanese

Another dog that does great with apartment living is the Havanese. They are highly intelligent, making them easily trainable. Furthermore, these dogs are full of energy and love the outdoors. This means that you will need to make sure your Havanese can get outside to let run off some of their energy. At the same time, because of their smaller size, they will get a lot of running done indoors.

Like most other dogs, the Havanese's coat needs to be brushed a couple of times a week. They don't tend to shed, so are great hypoallergenic dogs. When training a Havanese, positive reinforcement is your best procedure. They are sensitive dogs and quickly understand your tone of voice. Once they realize the tones, they will judge how you feel about them and their actions by your tone.

Shetland Sheepdog

The Shetland Sheepdog is a competitive dog. They enjoy dog sports, are playful, and affectionate. They love learning new behaviors as it keeps their mind and body active. If you have a Shetland Sheepdog, you should enroll them in obedience or agility classes because they will thrive with this additional training. Like the German Shepherds, Shetland Sheepdogs want to please people. This causes them to work hard to learn new tasks. Because of their high intelligence, they learn and retain new information easily.

Shetland Sheepdogs are great to have in the country but can adapt to apartment living. However, they are extremely active and can grow about 13 to 16 inches at the shoulder when standing. They are sensitive dogs and don't tend to bark unless they feel danger or a stranger is approaching. Therefore, they make great watchdogs.

The Papillon Club

The Papillon is an active little dog who needs to be kept busy. They don't get much taller than 11 inches at the shoulder. If you choose this breed, you will want to think about an active training schedule. The training you give the Papillon should mentally motivate them as well as physically. They are curious dogs and quick to learn.

Unlike some breeds, the Papillon learns from each experience, and they remember everything. Therefore, you need to maintain consistency as a trainer. If you are not consistent, your Papillon is not going to thrive. They will become confused and seem like they lack skills, including potty training.

Border Terrier

Border Terriers are a smaller dog that can easily adapt to any living situation. While they are not the most popular dogs, they are great dogs for families. They love children, love to learn, and great dogs for training. Because border terriers enjoy agility classes, they are perfect dogs for easy training.

Australian Shepherd

The Australian Shepherd is a medium-sized dog, which makes apartment living a bit of a challenge for them. They don't like to spend a lot of time indoors. Australian Shepherds thrive best when they have a large area to run around, play, and learn.

Australian Shepherds are known for their herding instincts. For example, many people will train them for herding sheep or cows from one pasture to the other. However, some owners have also found their dogs trying to herd a group of children. The dogs will not hurt the children, but younger children can become confused and frightened when an Australian Shepherd is trying to herd them into a designated area.

The best home for an Australian shepherd is with someone who can devote a lot of time in training them. They are dogs that need a lot of love and a firm voice to keep them in line. They can lose sight of what they are supposed to do because of their boundless energy.

While some breeds are fine with training starting later in life, Australian Shepherds need to start as puppies. They are happy when their owners take time to train them and can work off much of their energy.

Golden Retriever

The Golden Retriever is one of the most popular breeds. They are known for their beauty and hunting skills. They are also used as seeing-eye dogs and for search and rescue missions. They are willing to learn and enjoy competitive events. A Golden Retriever is a dog that you would enroll in an obedience class because they enjoy this type of setting.

Golden Retrievers are family dogs that are eager to please. They are highly intelligent, making them easily trainable. They have a happy and playful approach to life, making them energetic and powerful dogs for outdoor activities. Because they are a medium-sized dog, they don't enjoy apartment life well. However, if you make sure they exercise daily and take them outside, they will do fine in an apartment.

Where to Go?

Now that you are briefed on some of the most trainable dog breeds, you might be asking yourself where to go. Thanks to the internet and social media, the places you can find a dog of your choosing is right at your fingertips.

Adoption

One of the first places to go is a local animal shelter or rescue group. There are a variety of dogs you can find in a shelter. Some places are mainly for purebreds, but most take in any dogs that need a place to stay until they are adopted.

One of the biggest challenges for people when adopting a dog is the unknown history and how to make the dog comfortable in their new home. To help your dog adjust to their new life, here are a few tips to follow.

- *Don't expect too much right away.* The new member of your family needs time to adjust. You shouldn't start to train them right away. Your main focus for the first few days to a couple of weeks is making them comfortable.
- *Understand they have emotional trauma.* Most dogs that come from a shelter have some type of emotional trauma, and you need to help them heal. This is going to take time. The dog was abandoned or got lost. They don't understand what happened to their companion, who they still love and miss. Plus, their time in the shelter can cause emotional damage, even if they were treated well. Be patient as you work through this trauma with them; it will take years to repair. Unfortunately, some trauma will remain with the dog.
- *The dog might act differently in the shelter.* If you come across a dog you fall in love with, but they act scared and bark a lot doesn't mean they will act this way in your home. The shelter makes them anxious, causing them to react in this way. Don't believe they will always act in that way. Once you bring them home and they become comfortable, the dog will relax and act typical for their breed.
- *Start a routine right away.* One way to help the dog feel comfortable in their new environment is to establish a routine immediately. This doesn't mean you start training them right away. Instead, you will feed them at a regular time, show them where their water is, place their bed in a certain area, go to bed at a specific time, get up at a regular time, and play with them. If you struggle with routines yourself, get a planner and what down what to do when with your new dog. Be as consistent as possible because they will start to trust what is happening next. If their routine changes, it can make them anxious.

Pet Store

I won't deny this—there is a lot of debate about pet stores. Some pet stores do receive dogs from puppy mills, but not all. Do thorough research on your pet store before you decide to get a dog there. Go to visit the possible animal a couple of times and get to know the employees. While they can't hold the dog you are interested in for you, knowing the background of the store will help you understand the dog better.

Some pet store will advertise that they are "puppy friendly," meaning they don't use puppy mills as a resource. While this can be true, remember anything can be said through advertising. Through research, you should find out where the puppies come from as some stores do take homeless dogs.

Responsible Breeders

Some people breed dogs. These dogs are usually purebred and well taken care of. Of course, you will still want to get to know the breeders. If you can, talk to other people who have gotten puppies from them. Interview the breeders and get to know the breeders. Ask them why they decided to breed and how long they've had the parents.

Social Media

There are a lot of social media sites that will sell puppies. Usually, the people have a dog that became pregnant and they can't care for all the puppies. Sometimes they are giving away the dogs and sometimes asking for money. Typically, they don't ask for a lot of money, but it depends on the breed. If you feel that they are asking too much for the dog, do a little research. They might be part of a puppy mill.

You always want to be cautious of online social media sales. While most occur locally, there are people willing to ship the dog if you send the money to them. This can easily be a scam, just like any smaller business stating the same thing. Always do your research if you don't know the people.

Tips to Keep in Mind Before Bringing Your Dog Home

It is always a bit challenging for some of us to say we are searching for our "best" dog. If you love dogs and want to take one home, you are going to be fine with getting a free puppy out of a cardboard box from your neighbor's home. Of course, there is nothing wrong with this. But, for many people, finding the best dog is necessary for several reasons. Whether you are looking at your neighbor's new litter of puppies or thinking of heading to the animal shelter, here are a few tips to help you along the way.

Be Cautious of Internet Sites

You are searching online for your Australian Shepherd. You know there isn't an Australian Shepherd available locally. However, you notice this small business called "Worldwide Love For Dogs Rescue." As you read about the newly established company, you come to learn that they will talk to you through video chat where you can see your potential dog. You can also receive email information about your dog's personality before making the final purchase. Then, once you send them the money, they will transport your dog to you. Sounds like a pretty good deal if you find the perfect Australian Shepherd for your home, right? This can be a good deal, but it can also be a scam.

One factor to consider is what do you really know about the company? You should always read up on any reviews, whether they are positive or negative. Do a little research on the company. Are they establish as an LLC or another type of organization? You can often find this information online. Do they keep their business clean?

Another factor to consider is how well they treat their dogs. People can easily tell you how they treat animals in a way that makes you believe they take care of their dogs. But you never truly know until you see the dog or do a little research on the company.

It is always a good idea to do a background check on the person you are thinking of doing business with. While you might have to pay for a background check, there is information you can find for free through research. Remember, people can say anything in advertising. No one is going to tell you that they have 125 dogs caged in unsanitary conditions because they are running a puppy mill or into dog fighting.

Visit the Dog on Site

Another reason many people will caution anyone against getting a dog through an online store where they are shipped is because visiting the dog on site is 100% helpful. There are hardly any downsides to visiting the dog before you decide to take them home. The only real downside is you are going to fall in love with the dog and have to wait until any paperwork and your background check clears or you will fall in love with all the dogs and can't bring them all home!

Seeing where the dog sleeps and temporarily lives will allow you to understand the dog. You will see firsthand how well they eat and interact with other people and animals. You will also make the dog more comfortable when you bring them home. Dogs can be nervous when they are chosen as they don't understand what is happening. While they might seem excited, like any animal, they will be cautious of their new surroundings and home. Meeting the dog in their home can help them feel more comfortable during transportation and their new living situation.

Poor Breeding Practices and Living Conditions

Visiting the site will help you see how well the dog's temporary residence is taken care of. If you walk into a building and see dozens or hundreds of dogs, some that are kept in the same cage, the building smells bad, and the dogs seems scared, ungroomed, thin, and not cared for (i.e., water might be dirty or water bowl empty) you might be in the middle of a puppy mill situation. It's important to note that animal shelters have standards they need to ensure they follow. Puppy mills are typically illegal and aren't kept up to standards.

If you ever find yourself in this situation, the best step to take is to try to get some background information. Are they running an animal shelter that is registered as a business? Closely check the conditions of the dogs. If you notice they don't give you a lot of information or don't answer your questions, contact the police department or a local animal welfare agency. It's always better to ensure that the dogs are well taken care of than to allow a puppy mill to continue.

While you might want to get one of the puppies to bring home for a better life, you need to be aware that dogs from puppy mills are often sick. They can carry diseases from the unsanitary conditions and need to be taken to a veterinarian immediately.

Chapter 2: E-Collar Basics

When you look at the instructions for an e-collar, you may feel confident you have all the information you need. Unfortunately, you are wrong. While companies give you detailed instructions when you purchase their e-collar, it can still be confusing for many reasons. First, each breed is going to react differently to the e-collar. This will happen because of their personality and how trainable they are. For example, some dogs will struggle with the e-collar because they are more independent and don't care to please their owners like other breeds. Second, each button is going to make the dog feel a different way, making your dog react in different ways to the e-collar. This can throw you off as you fear you might be doing something wrong.

Don't allow the way your dog reacts to the buttons throw you off. This is going to throw your dog off as well. Dogs can sense how you feel. For example, if you push a button on the e-collar and you notice them take a step back, you start to feel worried you hurt them. Your dog will sense this, causing them more uneasiness with the e-collar and training in general.

What Is the E-Collar?

A little over 30 years ago, the first electric dog collar made an appearance. Initially used to train hunting dogs, people became excited and worried about how the collar affected their dog. Some people felt they are harmful while other people believed they are a great training tool. The first e-collar only delivered one stimulation level. Today, there are three stimulation levels–high, medium, and low–and various levels of intensity within these stimulation levels for many e-collars. It is safe to say that over the last three decades, e-collars are safer, made for all types of dogs, and a perfect training tool.

The e-collar is a collar that has a radio receiver attached to it. The radio will receive a signal from the transmitter. This signal will give your dog a small electrical shock, similar to how you feel static electricity. This startles your dog enough to know that what they are doing is not appropriate. Over time, you will train your dog through the e-collar to know what behavior is right and wrong.

Most people choose e-collars because they don't want to train their dog with a leash. Another reason people choose e-collars is that they can train their dog from a small distance. For example, you are at the park with your dog when they start running off. Instead of yelling at your dog, you press a button. This alerts your dog to stop running and come back to you. Of course, at this point you have trained your dog to respond to the shock by looking at you to see what you want. For instance, you will motion them to come back. If you want them to sit, you will motion them to sit.

You can purchase a e-collar in most stores or online. If you've never worked with an e-collar, you might want to go to your local pet store. The employees there will help you understand how the e-collar works and answer any questions you have.

Types of E-Collars

When searching for the best e-collar for your dog, you need to think of why you're training. E-collar aren't always used for basic training, but some are focused for yard use. There are also e-collars for hunting, no barking, and working dogs.

Yard Training

Yard training e-collars is mainly for your pets, but some can be used for other training purposes. They are some of the most common e-collars and give you a lot of options.

- *EZ 900 Easy Educator* is an e-collar that focuses on safety for your dog. Its boosting level reaches to 60 and stimulation level to 100.

Book 1 - E-Collar Training Step-By-Step

It's considered the most humane e-collar on the market. This e-collar is a smaller size, making it great for any size dog. Many people who purchased this product say you won't see the typical head jerking coming from your dog. You can use the EZ 900 up to ½ a mile and can be used for hunting and working training as well. Another benefit of this collar is it's one of the cheaper ones on the market.

- *ET-302 Zen Educator* is another humane collar that can reach about a ½ mile. The stimulation level reaches to 100 and holds boosting levels from 1 to 60. It includes a collar to help you find your dog in the dark and is waterproof. This collar is sized for any dog and has two receivers, but is specifically for yard use.
- *ET-402 Educator* has stimulation levels from 1 to 100 and boosting levels from 1 to 60. This e-collar can reach up to ¾ of a mile and is great for emergency situations. It's waterproof and includes a light. Any size dog can wear this collar, but it's specific to yard training.
- The *FT-330 Finger Trainer Educator* is one of the newer models that can reach up to ½ a mile. It's a smaller collar, but holds enough stimulation for any sized dog. This collar holds most of the same specifics of the other collars, but includes a remote finger button.
- *ME-300 Micro Educator* works best for smaller dogs. It's a smaller version of the RX-090 Mini Educator Receiver but holds the same features. Its range is a ⅓ mile and charges quicker than other e-collars.

Hunting Training

While a few yard training e-collars are used for hunting as well, many professionals feel it's better to get a specific hunting collar.

- *UL-1200 Upland Dog Trainer* is one of the cheaper hunting e-collars, but it's also older than some models. Its stimulation and boosting levels are the same as other e-collars. You won't want to use this e-collar on smaller dogs because of its size. One benefit of this e-collar is it can be used for yard and working training as well.

- *WF-1202 Waterfowler* is a perfect hunting training e-collar. It's the best on the market and can reach up to one mile. While this e-collar is useful for any type of training, it's only for medium and large dogs.

Work Training

Work training e-collars are perfect for dogs with certain jobs and performances. They are sometimes used in seeing-eye dog training or dogs preparing for dog shows.

- *ET-802 Dog Remote Trainer* is a newer version of the ET-800 Dog Remote Trainer. It holds the same type of stimulation and boosting levels as other e-collars. This e-collar is usable for any type of training, but only used on medium or large dogs. It's weight and size is too much for a smaller dog to handle. It can reach up to one mile.
- The *K9-402 K9 Handler* can reach up to ¾ of a mile. It's one of the smallest working dog training e-collars, making it perfect for any sized dog. You can also use this e-collar for yard or hunting training.
- *Pro Educator 900* is an older e-collar built with safety in mind. It's smaller in size, so perfect for any type of dog. It reaches up to ½ a mile and is also great to use as a no-bark collar.

No-Bark E-Collar

There aren't a lot of e-collars specific for no barking as you can typically use any type of e-collar for this training. However, the main no bark e-collar is the *BP-504 Anti Bark Collar*. It's newer on the market, has nine stimulation levels and three sensitivity levels. This e-collar is great for no barking because it comes with a warning sound your dog will quickly understand. This e-collar should not be used for any dog under five pounds.

E-Collar Accessories

Each e-collar will come with a manual and it's necessary pieces, such as the transmitter, battery charger, lanyard, contact point tool, e-collar, strap, and extra contact points. You can also purchase various accessories to go with your e-collar. For instance, many people like to purchase another contact point tool as it allows for more than one person to control the dog. As long as you are both on the same page, this won't confuse your dog. You just want to ensure that you are both aware of what type of training you're focusing on. Other accessories include:

- A carrying case for all your dog training tools. Some people don't keep the e-collar on their dog as it's only for training purposes. For example, if you are training your dog to hunt, you don't need to have the e-collar on your dog when you're at home.
- A dummy e-collar. Some people purchase this accessory because it makes your dog believe the collar is still on when it's charging. If you are not constantly training your dog, this isn't necessary unless you want to test your dog on their training.
- The Educator Gear Keeper is great for anyone who is training their dog on the go. You can attach this device to your belt loop or purse to give you quick access. It allows you to put away the transmitter instead of keeping it in your hand constantly.
- Educator bungee straps are popular because they come in a variety of colors. They come in 33-inch and 37-inch lengths but can be cut down to size. They are also extremely comfortable for your dog and easy to clean. Unfortunately, they are not the best for smaller dogs.

How to Use an E-Collar

E-collars are not meant to be used as a source of punishment. They are meant to deter the dog from unsafe and negative behavior. The purpose is an e-collar will train the dog that their negative behavior gives them an uncomfortable shock. The dog will remember this sock and stop the behavior. This theory will only work if you don't allow your dog to know you control the collar and you allow them to become comfortable with the collar.

Parts and Terminology

- **Stimulation level.** The best e-collars to get are the ones that have at least 100 levels of stimulation. The more levels you have, the easier it is to train and select the best shock for your dog.
- **Power Output.** Do not assume that larger dogs need a higher stimulation level. Dog trainers often find lower stimulation levels are great for dogs over 100 pounds. At the same time, your smaller dog might need a higher level. It's difficult to find the "sweet spot" for stimulation, but by paying attention to your dog's reaction and talking to a professional trainer, if necessary, it's 100% possible.
- **Vibrate or tone.** Find an e-collar that allows you to emit a collar vibration or audible tone. These are used as warning signs that the shock is going to happen. What typically happens is that your dog will start to notice the tone or vibration and stop the behavior immediately, so a shock isn't needed.
- **Distance.** It is helpful to get an e-collar with considerable distance. A ¼ of a mile is minimum and a mile is usually maximum. The average length is ½ mile and perfect for house training. If you are more interested in hunting, you will find e-collars with one-mile distance.

Step One: Read the Instructions

Every e-collar should come with instructions. If you purchase a e-collar from a thrift store, garage sale, or a third party you can easily find instructions online. It is essential that you follow the instructions given to you. You always want to read the instructions before you put the e-collar together and attach it to your dog.

Step Two: Place the Batteries Into the Transmitter

Your transmitter is your remote. You will use this to control the e-collar, telling it when to give your dog a little shock. You want to ensure that the e-collar works with the transmitter and the levels aren't too high for your dog. You can do this by testing it on yourself first.

It is always best to put the e-collar on its lowest setting to start, even on larger dogs. Many people feel that larger dogs need a higher setting to feel the shock. This isn't necessarily true. Dogs are extremely sensitive to the shock they are given, no matter what their size. While a small dog will feel a stronger shock from a larger e-collar, bigger dogs can feel the same type of shock as a small dog from a smaller e-collar. Another reason to place the e-collar on the lowest setting is so you don't accidentally shock your dog when fitting the collar.

Step Three: Fit the E-Collar on Your Dog's Neck

With the e-collar turned off or on the lowest setting, fit the collar on your dog. Make sure that you can place your pointer and middle finger between the collar and your dog's neck. This is the general rule of thumb for any collar to ensure that it's not on too tight.

Be aware of e-collars that have small prongs. These have to touch the dog's neck, but you don't want to make it uncomfortable. You might need to adjust the e-collar a couple of times once you use it before you find the best fit.

One of the biggest mistakes is putting the e-collar too low. A dog's neck is shaped bigger at the base of their neck, causing the e-collar to move up the neck when they start playing. Here are some tips for fitting the e-collar:

- If your dog has thick fur, de-shed their neck first.
- Place the e-collar high on your dog's neck, but not at the top.
- Use two fingers to ensure it is a snug fit, but not too tight.
- Rotate the e-collar every couple of hours so your dog's neck doesn't get irritated by the collar.

Step Four: Let Your Dog Adjust to the E-Collar for a Week

While this step is optional, many professionals say it's best to not use the e-collar for a week. This will let your dog get used to the collar before it feels the first shock and not associate it with punishment. The point of the e-collar is to make the dog believe the negative behavior is causing the shock. If the dog knows it's the e-collar, the training can become difficult and they can try to escape from the collar.

Step Five: Start Using the E-Collar

Start with the lowest stimulation level and observe your dog's actions. If they react, such as moving their head or twitching their ears, keep the level low. If they don't react, upgrade the level. The dog should never whimper or try to run away because of a shock. If your dog makes noise like they are hurt, the shock is too strong and you need to lower the levels. If they're at the lowest level, you need to find a smaller e-collar for your dog.

Always be consistent when using the e-collar for training. For example, you don't want your dog jumping on the furniture, use the e-collar every time you see them do this. They might still do it when you aren't looking, but if you are consistent they will learn not to jump.

Step Six: Start with Commands Your Dog Understands

Your dog should already know basic training tricks, such as sitting. When first using the e-collar, start by saying commands they understand. If they don't respond, use the e-collar to get their attention. When the shock is received, repeat the command. Don't do this repeatedly over a few minutes as this can stress out your dog. Follow this step whenever you want your dog to sit, stay, or lay down. Every time your dog responds remember to praise them.

Step Seven: Control Your Dog's Negative Behavior

After your dog responds to basic commands, use the e-collar to control their negative behavior. For instance, if your dog jumps on company, use the e-collar every time you see them jump. While many people find their dog's growling or barking annoying, don't shock them immediately. Remember, this is how dogs communicate when something seems wrong to them. If they are barking because a stranger is coming toward your home, think about if shocking them is the best choice. Dogs are territorial and have natural protective instincts. You don't want to silence these instincts.

Do your best to ensure your dog doesn't see you using the transmitter. They are smart animals and will connect the shock to your transmitter. If they see you shocking them for their behavior, they will start distrusting you. You want your dog to believe the shock happens because of the negative behavior and not you.

E-Collar Safety

Don't Leave Your Dog Unattended with an E-Collar

Most people believe that you should have the e-collar on your dog, but this isn't necessarily safe. Your best step is to purchase the dummy e-collar and place this on your dog when you aren't training them, during the night, or when they are home alone. There is always a possibility that the e-collar will overcorrect or malfunction. While this is rare, it can harm your dog if it happens. Of course, because this is a rare occurrence, it is up to you.

Understand Your E-Collar

One of the best ways to practice safety with the e-collar is to ensure you understand everything about the e-collar. You have a vast amount of knowledge and know how to use your transmitter before you begin training. For example, you understand what all the buttons do and are prepared to follow your dog's reaction to know when the e-collar is at the right stimulation level.

Don't Use a Leash with the E-Collar

Using a leash with the e-collar can cause the front of the collar to push up against the dog's neck if the leash is pulled one way or another. If you shock them at the same time, this can harm them. You can use a harness that is attached to the leash, but don't attach any part of the leash or harness to the e-collar.

Don't Use the E-Collar When Your Dog is Swimming

While e-collars are waterproof, it's safer to keep them out of the water. If your dog is going to swim, remove the e-collar and don't replace it until their neck is dry. Following this rule not only secures your dog's safety but can make you e-collar last longer. When you need to wash the e-collar, always use warm and soapy water.

Benefits of an E-Collar

Fast Results

One of the challenges of training is that people don't have the time or patience to effectively train their dog. However, people who use the e-collar state it provides fast results and only takes a few shocks for the dog to stop the negative behavior. This works because before the dog feels the jolt, they will feel the warning vibration, associating this warning to the shock. Therefore, feeling the vibration makes them stop the behavior so you don't have to give them a shock.

Unfortunately, not everyone has had the quick success rates. Dogs that are stubborn tend to be harder to train. If you don't have an easily trainable canine, you shouldn't expect quick results.

Long Lasting Behavioral Changes

Professional trainers say e-collar training is the best way to change your dog's behavior, as long as it's used correctly. Studies have proved that dogs remember the behavior that brought on the shock better than a tone of voice or any other type of training technique. Hunting dogs trained not to go too close to the sheep received a shock. A year later, the dogs returned for another test and demonstrated hesitation when they started wandering too close to the sheep, proving they remember the shock when they got too close to the sheep. In fact, only one out of 114 dogs required a shock during the return test (Evans, 2018).

E-Collars Don't Take a Lot of Strength

When you use a leash for training, you need to have a tight grip and remain strong, especially with bigger dogs. They can easily pull a person around if you aren't strong enough to handle your dog. Strength isn't a concern with the e-collar because you use the transmitter to give the dog a warning noise or vibration and then a shock if their behavior doesn't change. Your dog can even be a ¼ of a mile away from you and still feel the warning and shock.

You Don't Stress Your Voice

Most dog owners know the stress your voice can feel when your dog doesn't listen to you, especially when their safety is concerned. For example, your dog escapes from their leash and runs into the street. Your automatic reaction is to yell at the top of your lungs for your dog to come back so they aren't hit by a driver. This can damage your vocal cords if you aren't careful or find yourself yelling often. With the e-collar, you don't need to say a word. You simply press the button and the e-collar does the rest.

Yelling or speaking to your dog angrily can cause them to become stressed, affecting them emotionally. They will become confused about what exactly they did wrong and worry about what will happen next. This poor communication between you and your dog can change with the e-collar. First, if you properly train them, they won't associate you to the e-collar. This will make them more comfortable with you. Second, your dog will receive clear communication in their current behavior by giving them a shock. They will come to understand that this behavior is unpleasant, making them stop.

E-Collars Allow for Easier Consistency

Training your dog with a leash or voice is difficult. Some people are afraid to train their dog in public using their voice for fear of causing a scene or judgment. The e-collar allows you to train your dog wherever you are without people realizing you are shocking your dog unless they see you push the button. If you do struggle with social anxiety and still worry what people will say or do when they learn you use an e-collar, you can purchase small transmitter to fit onto your belt or in your purse and push the button more discreetly. However, this should not become an issue because most people understand e-collars do not harm dogs.

Your Dog Receives Off-Leash Freedom

Leashes keep dogs in a certain area. While you can get a long leash, this can cause problems with your dog wrapping themselves around a tree or another object. Furthermore, if people aren't fully paying attention they can trip over the leash. E-collars will give your dog the freedom of running and playing without the tug of the leash. It is also safer as your dog can't tangle themselves up or trip anyone.

Common Myths About the E-Collar

The e-collar received a lot of criticism when it first became public. People automatically saw the e-collar as a harmful device for their companion. After all, who would want to shock their dog in order to teach them right from wrong? People who used e-collars were asked, "How would you feel if I shocked you when you did something wrong?" The harm an e-collar can cause a dog soon became a myth that still exists today.

The Shocks Harm Your Dog

Before I go too far into the world of e-collars, I want to tell you that the shocks your dog receives will not harm them. I love my dogs and have used e-collars for years. My dogs have never shown any signs of harm from the e-collars. Like you, I did a lot of research before I finally found the e-collar for my dogs. Since I purchased the e-collar, I will not go back to any other training technique. If my dogs ever showed signs of the collar hurting them, I would have stopped using it in a heartbeat. The shocks the dogs receive are similar to a static shock. While they are not fun to receive, they don't harm you.

If you take your dog to a training class that uses e-collars, they might have you try the e-collar yourself. I did this when I received the e-collar. I became surprised by the shock and felt more comfortable using it because it did not hurt me in any way. Therefore, I knew it wouldn't hurt my dogs.

Only Professional Trainers Can Use E-collars

Some people believe e-collars are difficult to use. While e-collars were difficult to use when they first came out, technology has changed. Today, they are easy to use. You don't need to be a professional trainer or talk to a professional to use e-collars.

The E-Collar Burns Your Dog

The e-collar does not burn the dog. There is no way that the e-collar can burn any dog because it does not get hot. The e-collar does not give off any type of heat when it gently shocks your dog.

The E-Collar Leaves Marks On the Dog's Neck

If you do see any marks from the e-collar, they are pressure marks. You have to rotate the e-collar on your dog at least every four hours. Furthermore, you should ensure you can place two fingers between your dog's collar and their neck. If you can't fit two fingers, the collar is too tight, making the e-collar uncomfortable for your companion.

Chapter 3: Your Dog and Their E-Collar

Before you choose an e-collar, you need to know why you want the training tool. For example, are you going to train your dog at home, for hunting, or to stop them from barking. The type of training goals you have depend on what e-collar you will get. At the same time, you need to remember your dog's size as not all e-collars fit well for smaller dogs.

Choosing the Best E-Collar

One of the toughest decisions you will make when e-collar training is what e-collar to buy for your dog. The list of e-collars in Chapter 2 should give you an idea of what to look for, depending on your training goals. For example, if you are going to house train, you want to focus on yard training e-collars.

The Fundamental Five

One way to help you choose the best e-collar is to follow these five tips.

1. *Choose an e-collar you can count on*

    Don't buy the cheapest e-collar for your training that you can find. You want to find an e-collar that is waterproof and has great reviews. Take time in looking for the best e-collar and check out online reviews on the product. Compare and contrast the reviews with different type of e-collars. This will help you narrow your decision a little more.

2. *Education is always number one*

Book 1 - E-Collar Training Step-By-Step

Some of the best e-collars on the market will not only come with instructions but also training videos. You always want to focus on e-collars that show they are meant to train your dog and not punish your dog. Always remember, the e-collar is not a shortcut for training. It is a technique to help your dog understand that certain behavior is unwanted effectively and efficiently.

3. *Know your options when it comes to e-collars*

Do as much research as you feel necessary to reach a complete understanding of e-collars. This means you want to research your dog's breed and how they handle training, know that your dog's age is going to affect training, and understand that some dogs are stubborn and will require longer training. This doesn't mean you aren't doing your job as their trainer, it's part of their personality. Look for an e-collar that holds constant and nick options and allows you to switch between the two easily.

4. *Choose an e-collar you understand*

Some e-collars are more complicated to use than others. You don't want to pick an e-collar that will frustrate you. Pick an e-collar that doesn't have a lot of buttons as this means you need to look for the right button to press. By the time you find the button, the teaching moment is gone and your dog won't understand their previous action warranted the shock. They will relate the shock to their current action. Another method is to choose a small transmitter that you can fit in your pocket or clip to your belt for easy and quick access.

5. *Pick a collar only your dog can activate*

There are e-collars that other dogs can activate, especially if you are using a no bark collar. When your dog receives a shock that wasn't meant for them at the time, they will become confused and unsure of why that happened, especially if they were listening, laying down, or sleeping. The best e-collars to purchase are the ones that will only set off a shock if it comes from the dog wearing the collar.

Your Dog's Reaction to the E-Collar

I have mentioned a bit about what to watch out for before, but I want to take this time to thoroughly explain your dog's reaction to the e-collar.

First, the e-collar can distress your dog. This is always a possibility and one that can happen even before you shock your dog for the first time. One of the best tips when it comes to knowing how your dog is going to react to the e-collar is to understand your dog before you even buy an e-collar. For example, if your dog came from a rescue shelter and is easily frightened, the possibility of your dog becoming scared by the e-collar is high. This isn't going to help the training situation, especially when they receive a shock.

Another way to know how your dog might react to the e-collar is to place a different collar on your dog. This is a good idea for dogs that don't have a collar. You will understand how they react to a regular collar and if you will need to speak to a trainer for the best advice on how to introduce your dog to an e-collar.

Unintended Reactions

It is always possible that your dog is going to have reactions you didn't expect to the e-collar. This can happen when they receive their first shock or throughout their training. You might believe you are training them not to leave the yard, when you are really training them to be afraid of someone. Let's look at a couple of case examples so you get a better idea of unintended reactions.

Case Example #1: Tobey and the Neighbor

Tobey is a one-year-old poodle that likes to run over to the neighbor when he sees them outside. While this typically wouldn't be a problem for Tobey's owners, they are concerned for his safety because the neighbor lives across the street. The neighbor is always kind enough to walk Tobey back home, but anything can happen when Tobey sees the neighbor and runs into the street.

Other than the neighbor, Tobey doesn't leave the yard. Therefore, his owners weren't sure that the e-collar would be the best choice to train Tobey not to run out into the street, but they decided it is the best options for Tobey's safety.

Every time they bring Tobey outside, they put the e-collar on him. Every time Tobey ran toward the street because he saw the neighbor, he received a shock. One day, the neighbor came over and as soon as they walked in the door, Tobey bit the neighbor on the leg.

The neighbor and Tobey's owners were shocked by his reaction. They knew he loved to visit the neighbor, so why did he bite them? After speaking to Tobey's trainer, his owners found out that Tobey didn't associate the road to the shock. Instead, he associated the neighbor to the shock. This happened because Tobey's goal was to see the neighbor when he received the shocks. He wasn't thinking about the street or that this was the real cause of the shocks from the collar.

Case Example #2: Donnie and Max

Donnie's parents gave him a seven-month-old puppy named Max for his 10th birthday. The family loved Max, but Donnie's mother worried whenever he took Max for a walk. While she always went with, Donnie held Max's leash. Because Max is an energetic puppy, he often pulled Donnie, especially when he saw another dog.

After talking to her husband, Donnie's parents decided to purchase an e-collar for Max. On their walks, Donnie's mother would shock Max every time he started to pull Donnie. While she felt this worked well at first, she started to notice Max become anxious and aggressive whenever he saw another dog.

One day, Donnie and his mom were walking Max when they met family friends walking their dog. As Donnie and his mom started walking toward their friends, Max stopped and refused to get any closer. Because the family friends felt Max was tired or simply being a puppy, they continued to run up to Donnie and his mother. However, the closer they got, the harder Max tried to go the other way.

"I don't understand why Max started this behavior," Donnie's mother said to Max's trainer. "He always liked meeting dogs before. Now he becomes anxious, growls, and I am afraid he is going to attack another dog one day. What happened?"

"Max didn't associate the shock to pulling Donnie," Max's trainer stated. "He associated another dog with the shock. Therefore, Max believes that if another dog gets too close, he is going to receive a shock."

Dogs are not mind readers. They don't always understand our goals when it comes to their training, such as the case examples show. Both dogs would have received more efficient training without the shock collars. Because Tobey didn't run out into the road unless he saw the neighbor, his owners could have used positive reinforcement to sit on the sidewalk and wait for the neighbor to come over. The same goes for Max, positive reinforcement to not pull Donnie when another dog is spotted would have worked better. You always want to take time to think about what your dog's goals are when they are taking part in unwanted behavior. If they are trying to meet another dog or person, it is best not to shock them because they will associate their mission to the shock.

What If Your Dog Is Scared of the E-Collar

One of the challenges to e-collars you can have is your dog being afraid of the e-collar. This can happen due to two main reasons. First, the dog is frightened of any collar going around his neck. This can happen because of past experiences or because your dog is naturally more fearful than other dogs. Depending on the reason will depend on how you handle the situation. For example, a dog who is naturally fearful is going to overcome the fear easier than a dog who has a negative past experiences with a collar.

There is a lot of debate when it comes to using an e-collar on a fearful dog. Most people feel you should never use an e-collar on a dog that exhibits a lot of fear because their fear will increase. Other people believe this is one of the many myths about e-collars and if you train the dog correctly, you don't have to worry about their fear.

First, you need to understand where the fear comes from. If they are afraid of collars in general, you want them to get comfortable with a different collar and then work toward the e-collar. Even if you decide to leave the e-collar on for a week or two without delivering any shocks, you should start with a regular collar. Another tip is to introduce the regular collar slowly. For example, let them smell the collar and then sit it aside. Later, bring the collar back to them and set it down next to them. Watch your dog's reaction and if they start to show fear, take the collar away. Go slowly, step by step, until you can place the collar around your dog's neck without them trying to break free from it. Once they wear the collar for a week or two without trying to get rid of the collar, then work them into the e-collar. You might have to go a bit slower with introducing this collar because it is a different collar than the first one, but it will go smoothly after a while.

There is a possibility that your dog will not become comfortable with an e-collar. If your dog is from a shelter, a rescue, or from a pet store they may have terrible memories of collars. This trauma is something that they will never truly get over, just like any trauma a human can face. Always remember, there are other ways to train your dog that don't include the e-collar. Your dog's mental health is more important than any type of training.

Some dogs are fine when you put the e-collar on them and they get used to the collar. However, after you shock them once or twice, they become scared. For example, Robbie is a beagle that is learning how to hunt. His owner put an e-collar on him two weeks ago and has used it twice to get Robbie to come back once he has found the duck. One day, Robbie's owner tones him as a warning the shock is about to occur if he doesn't come back. Instead of heading back, Robbie drops to the ground and doesn't move. Concerned, his owner goes to him and notices he is scared. Robbie looks at his owner and whimpers a little.

A week later, Robbie and his owner are at a training class. His owner explains what happened and asked the trainer what to do. The trainer tell Robbie's owner that he is afraid of the shock. "This happens with some dogs. They understand the tone means a shock can happen and freeze because they don't want the shock and don't know how to protect themselves from it. Beagles aren't the easiest dogs to train with the e-collar for this reason, many tend to become afraid and freeze. Tonight, we can talk about other training methods instead of the e-collar. It's best that you don't use the e-collar unless you absolutely need to."

Your dog can have a lot of reactions to the e-collar. Some dogs are known to become more aggressive when they are shocked. If your dog has a reaction that shows fear or aggression, you should stop using the e-collar immediately. Your dog will respond better to a reward system and leash training. I would start by placing your dog into a training class and discussing any other training techniques with a professional.

The Story of Frankie and her E-Collar

Frankie is a black lab who was taken in by her owners, Amirah and Thomas, because she couldn't become a show dog. She was born without any fur and while her first owners took her to the veterinarian and gave her medicine, fur never grew on the tip of her nose. Thomas, who worked with Frankie's initial owners, learned they were going to send Frankie to the shelter because they only keep show dogs to train and then sell. That's when Thomas said he would buy Frankie for his wife.

When Thomas brought Frankie home, she was seven months old. The couple had read up on how to raise a black lab because they knew she had

Book 1 - E-Collar Training Step-By-Step

natural hunting instincts, but they were not hunters.

Amirah and Thomas started the basic training with Frankie, but had a problem with her getting into the garbage. She would get into the garbage on the side of the garage, leaving it all over. Even when the couple caught her, they could barely pull her away from the garbage.

Worried about their dog's health from what she could eat in the garbage and tired of picking up trash all around their yard, they looked into other training methods. After researching the e-collar, they decided to give it a try.

Amirah, who works from home, spent her days training Frankie and followed the e-collar instructions. She let Frankie wear the collar for a week before she started to shock her. On the day, Amirah decided to start using the collar, Frankie walked toward the garbage. Once Frankie placed her two paws on the plastic garbage can, Amirah gave the warning and then a shock. Frankie moved her head a bit, but continued to go into the garbage. Cautiously, because Amirah didn't want to hurt Frankie, she gave the dog another warning and then a shock. This time Frankie barely moved.

Once Amirah got Frankie away from the garbage using her old tricks, she brought her dog in and adjusted to e-collar settings to a little higher shock. Later that night, Frankie got back into the garbage. Amirah gave her a warning and then shock, causing Frankie to jump back from the garbage.

When Amirah and Frankie came in, she told her husband what happened. They immediately called friends who use e-collars on their dogs and told them about it. "I would say that the level might be too strong for Frankie. She notices the shock at the first level you had her at, so I would move it back down. It sounds like you will need to give her the warning and shock before she gets to the garbage. Her mission isn't the garbage can, it is what is inside of the garbage can. Therefore, I don't think she will become frightened or aggressive toward garbage cans if she gets near one. Once she gets into the garbage, she is too interested in the garbage to care too much about the shock. Black labs are hunters and very determined dogs, it is hard to remove them from anything they set their minds on.

Amirah followed her friend's advice and noticed it helped. Within a couple of weeks, Frankie would walk toward the garbage, but then hesitate. The moment she started to hesitate, Amirah watched her dog closely. If Frankie took another step closer to the garbage, she gave the warning. If Frankie walked away, nothing happened. The dog wouldn't hear the warning sound. Soon, Frankie didn't worry about the garbage by the side of the garage.

Chapter 4: What You Need to Know Before Training Begins

Before you start training, you might feel overwhelmed about training your companion. At this point, it is important to remember that your dog and can sense your emotions. Therefore, if you are feeling anxious, they are going to start to feel the same way. They will associate their anxious feeling with training. This will interfere with their training and make their experience harder on you and your dog.

General Training Tips for Dogs at Any Age

There are tons of tips that you can use when training your dog. The most important factor to remember is you need to work with your dog. You need to do what is best for yourself and your dog. If you don't work together, the training system can easily fall apart. Below are several other training tips to help you and your dog thrive through the training experience.

Know Your House Rules

It's important that you don't just come up with a rule on the whim. Before you bring your puppy home and start to train, you need to make sure you know your house rules. You can do this by asking yourself basic questions. Do you want your dog laying on the couch? Are you going to let them freely eat or will you only feed them during certain times of the day? Are they going to have one area of the home they can stay in, run around the home, or will your dog remain outside?

Think about your dog's safety around your home. If you are getting a puppy, they can get into everything like toddlers do. It never hurts to ensure they can't open your cupboards with the cleaning products or get into your garbage.

No matter how hard you try, you won't come up with every single house rule right away. There are a lot of rules that you will think of once your dog comes into your home. For example, you might feel it's not okay to let them sleep with you on the bed, but have a change of heart once you bring them home. This is fine, I have done this myself, the key is you need to be consistent with your house rules. Don't change them once you have started training your dog because you "give up" on training or are "too tired to care." Inconsistency is going to confuse your dog and start to destroy trust.

Always Be Consistent

Consistency is one of the most important steps when you are training your dog. Consistency is one of the best ways they will learn. Furthermore, they will learn quickly, trust you, and know good vs bad behavior. Your dog wants to please you. Nearly every breed works on pleasing their owners. They don't want to do something wrong, but it is going to happen. It happens to everyone –human and animal–and is a part of the learning process. The more consistent you are, the easier your dog will adjust to their new home and rules.

Unfortunately, people struggle with consistency for many reasons. For example, you aren't home during the day, meaning you don't know what your dog does. When you come home and find they got into the garbage or tore up the pillow, you can't get after them at that moment because they won't understand. You always need to catch your animals in the act when teaching them what is right and wrong.

Some people aren't consistent because they have busy days and are tired when they come home. They tend to become relaxed with their dogs and let them do what they want because they don't want to train their companions. While I do understand how tired one can get at the end of the day, it is essential that you don't let yourself fall into this thinking trap.

That's all it is—a thinking trap. When you decide to bring home a dog and train them, it is your responsibility to keep up on this training, no matter how tired you are. Don't relax on your training because the one who is going to suffer from that is your dog.

Here are some tips to help you focus on consistency in your training:

1. **Keep your daily routines.** Set a schedule before you bring your new family member home and stick to it as much as possible. Of course, you will have emergency situations and something might change here and there. But, the more consistent you are with the schedule, the easier consistency is with training. Another reason to keep your daily routine is because it teaches the dog that these are normal parts of their day. Have you ever taken a minute to think about how stressful life can be for a dog? They have a lot of situations that can cause them anxiety, such as walking in busy traffic or being home alone for eight hours a day. A routine will help your dog feel more comfortable about their day.

2. **Be consistent with your cues.** Your dog is about to get into the garbage can on Sunday morning, so you say in a stern voice, "No." Your dog backs away and goes to play. The next morning, your dog is going to do the same thing and you tell them, "No no" in a lighter voice. Your dog continues to go toward the garbage, which is when you say more firmly "No," and your dog backs away. The problem with this example is you are not consistent. If you say "no" in a firm voice once, you need to do that every time. Don't add another "no" or change your tone of voice as this will confuse your dog, which is why they kept going toward the garbage. The same goes with any nonverbal cues.

3. **Keep your words simple.** If you are training your dog to come to you by saying the following phrases, "come here," "come," and "come now," you are confusing your dog because they don't mean the same thing. Dogs listen to every word you tell them, and they will understand something different when they hear "here" and

"now." Therefore, if you want your dog to come to you, simply use the word "come."

4. **Don't do all your daily training at once.** Training your dog for 20 minutes in one segment can become too much. They aren't going to remember everything, they will become confused, and they are going to get tired and annoyed. Your dog's attention span is similar to a toddlers–it is very short. The best option is to do any type of training in 2 - 3 minutes segments throughout the day.

5. **Everyone needs to be on the same page.** This can get a bit challenging if you have younger children who want to train, but everyone in the household should understand the house rules and train the dog in the same way. This can also be a challenge when you have company over. For example, you are training your dog not to jump on people, but your friend rewards the dog with petting and acting excited when your dog jumps on them. This will cause your training to take a step back. It is up to you to explain your training to your friend so they can help your dog understand that jumping on people is not acceptable. With everyone, including guests, being on the same page, your dog will quickly learn the house rules and stick with them.

Stay Healthy

Training is stressful, for both you and your dog. One of the best ways to give yourself the energy to train and keep your mind clear is to stay healthy. At the same time, you need to keep your dog healthy so they will have the best training experiences.

Making sure both you and your dog eat healthy is a great start to staying healthy. For dogs, you need to buy food for them that is nutritious, meaning you want to stick to their diet that is natural for them. This food will be a little more expensive than other food, but it will keep your dog healthy and strong. You want food that has a lot of protein, calcium, active enzymes, essential amino acids, and fatty acids. With these nutrients, your dog is more alert and pay more attention to what you are saying, your tone of voice, and your actions.

One factor you need to keep in mind with your dog's diet is transitioning. If you find you aren't feeding them the best balanced diet and find a different blend of food, such as Primal, you want to transition them to the new food. You want to start things off slow and steady, especially in the first week.

For the first two to three days, mix their regular food with the new food. At this point, you will have about ¾ their old food and ¼ cup Primal food. Watch your dog when they are eating to see how they react to the Primal food. If there are no problems and they eat it, add more Primal food and less old food on the fourth day by giving ½ old food and ½ Primal food.

It is possible that your dog can start having some bowel stress or gastrointestinal problems. When this happens, simply add one tablespoon of Goat Milk to their meal. Mix it well to ensure they get all the Goat Milk possible.

Once the seventh day rolls around, you can decrease the amount of old food to ¼ and increase the Primal formula to ¾. Then, on day ten, you will stop giving your dog any of their old and food strictly give them Primal food. While their system should be used to the Primal formula, it's always a good idea to monitor your dog's bowel movements for a few more days.

While you are working on your dog's diet, you can switch your diet as well. For instance, you might cut out sugars or eat food higher in healthy fats and lower in carbohydrates. Just like you do for your dog, introduce your new diet slowly and you will find that by day ten, both you and your dog are more alert and ready for the best training experience possible.

Getting enough sleep is another factor in staying healthy. It isn't always easy to make sure your dog gets the sleep they need, but they tend to do this pretty well. If you notice your dog struggling to sleep in their bed or in the area they are supposed to sleep in, do a little research to find out why. Another sign that your dog isn't sleeping well is they will find a different spot to sleep instead of lay in their bed. If you find your dog in a certain spot every morning, allow them to fall asleep in this spot and see if they remain their all night. Dogs will naturally get up throughout the night to stretch, check on you, or get some water. However, if they have a comfortable place to sleep, they will go straight back into that location until their day is supposed to start.

Be Patient

It is going to take time to train your dog and if they came from a shelter or have previous bad experiences, it will be harder. Dogs who come from abusive situations or are abandoned fear that these experiences are going to happen again. Like humans, they don't want to go through the physical and emotional pain it causes them. They want to feel loved in secure in their new home, but it is hard for them to trust you.

The best key to help your dog heal from any previous trauma is to be patient. Understand that your dog is emotionally and mentally hurting and you need to maintain a calm and trusting environment to help them through their emotions.

You also need to understand that some dogs will always have emotional and mental scars from previous trauma. They are similar to humans in this way, but aren't able to work through their emotions and mental scarring through therapy like humans can.

If you know or believe that your new family member was abused, here are some helpful tips to help your dog overcome their internal battles.

1. Get down to their level. Don't stand up and talk to your dog if they are afraid as this will make them feel inferior and increase their fear of you. Getting down to their level makes them feel equal to you.

2. Off them a treat. This isn't something everyone will do, but treats always make dogs feel better. Think about how often you want to

have a treat after a bad day. Dogs feel the same way. Plus, giving them a treat reminds them that you love and care about them, immediately improving their mental state.

3. Make sure your dog has a safe place. You will start to know when your dog is uncomfortable or afraid through their reactions. When you notice your dog showing signs of this behavior, bring them to a secure place. This could be their area in the house, garage, or any place they feel the most comfortable. If you aren't in their secure place at home, take them out of the environment and spend quality time with them.

4. Don't forget about a pet behaviorist. It's important that you don't give up on your dog. Doing this will only increase their emotional and psychological trauma. If you find that you are having trouble handling your pet's behavior because of their past abuse or abandonment, bring them to a pet behaviorist. They will help you understand where your pet is coming from and give you ideas on how to help your pet overcome their trauma.

Reward Good Behavior

As a trainer, you don't want to get in the habit of shocking your dog with unwanted behavior. For success, you need to reward good behavior, including when your dog backs away from the unwanted behavior. For example, you are outside and training your dog not to go into the chicken coop. You have a curious puppy, and they often find the noises coming from the chicken coop interesting. Every time they pass the gate into the chicken coop, you give your dog a warning before a shock. Each time you see your dog, you ask them to "come" and wave your hand. They follow this direction and you give them a treat for coming to you when you called.

If you are training your dog and know you will need to give them a lot of treats, don't focus on unhealthy or bigger treats. Get some healthy treats that will help boost your dog and not drag him down. Too many unhealthy treats can cause stomach issues and make your dog sick. It's also possible to give him a little piece of a bigger healthy treat each time he follows your direction while training.

Obedience Classes

No matter what age your dog is, you can always enroll them in an obedience class. In fact, this is a great way to make certain your dog is socializing and you are receiving the right help you need for training. There is a lot of information that goes into training, and it is difficult to simply start training your dog without research and advice.

There are some dog owners who feel they don't need to spend the money on obedience training because their dog is "good enough." While you have a great dog, there are many benefits that are included in obedience training.

1. **You will meet like-minded dog owners.** While you may have friends with dogs, this doesn't mean that take training as seriously as you do. When you take your dog to an obedience school, you will meet people who have some of the same ideas and goals for their dog that you do. You will find someone who can help you through the training process or find someone to talk to about you and your dog's failures and successes with training.

2. **You will expand your knowledge.** No matter how much research you do on training, obedience classes will expand your knowledge further. Your dog isn't the only one who will learn from the class.

3. **You will build your bond with your dog.** One of the best steps of obedience school is building your connection to your dog. This is a special time for you and your dog, and you will both feel it. Your dog will feel that you care about their general well-being and you will feel like you are doing everything you can to ensure your dog has a great training experience.

Book 1 - E-Collar Training Step-By-Step

What Happens in Obedience Classes?

Sometimes dog owners are weary about obedience classes because they feel only dogs with behavioral problems go to them. This is a myth as any dog at any age will benefit from an obedience class. Furthermore, any dog owner will benefit from a class.

Another reason dog owners are cautious about obedience classes is they don't understand what happens in them. They feel it is simply teaching your dog tricks, but this is only a part of it. Obedience classes help your dog to understand what their role is with you and within the world. They will not only learn the basic commands, but also social skills. You will begin to understand healthy behaviors from negative behaviors when it comes to your dog. In general, you will feel closer to your companion because you will understand what they are telling you when they act a certain way.

Your Dog's Age Matters

The age of your dog matters when you are trying to train them. While dogs can be trained at any age, it is a lot easier to train puppies than older dogs. Some people feel they have to get a puppy if they want to train a dog, yet don't have the energy or ability to take care of an energetic puppy. Fortunately, this is a myth and senior dogs can learn just as well as puppies.

When it comes down to the basics, a senior dog is going to learn just like a puppy. However, a senior dog may take more time to learn the tricks than puppies. Part of this is because a senior dog is set in their ways, just like human adults get set in their ways. Another part is because senior dogs are generally slower. They are going to take more time to reach their paw up to shake or lay down. As long as you are patient and consistent, you will find your training fits with dogs of any age.

If you recently got an older dog from a shelter, get whatever information you can about their background as this will help you understand how your dog is going to react to training. Another reason is because you want to try to learn what tricks older dogs already know. For example, were they taught to sit by the previous owner and how? While this might be impossible to learn, you can work with your dog, a pet behaviorist, and a trainer to get an idea of how they were trained. You can start by telling the dog to sit–if they sit repeatedly at this command, they understand. If they don't sit continuously, they may have received inconsistent training.

For older dogs, you need to understand if they have physical disabilities that will stop them from doing certain tricks. First, you should never train a senior dog to do highly active tricks, such as learning how to surf or riding a skateboard. Senior dogs don't have the same strong bones as puppies and can get hurt easily. A veterinarian will tell you if your new dog has any type of disabilities and what they can learn vs what is too much for them.

Before you take in an older dog, you need to make certain you are ready for the responsibility. A lot of people think puppies take more responsibility, but in reality, senior dogs can because of any health problems. For example, taking in a blind dog with arthritis means medication and a lot of trips to the veterinarian's office. You will also need to care for the dog just as much as a puppy. For example, a blind dog will need to be slowly taken around their environment, so they don't constantly run into things. They will usually walk slower and be more cautious because they are afraid of running into something.

Case Example: Alice the Trainer and Her Dogs

I have raised many dogs throughout my life. My father bought my first dog, a Pitbull puppy, when I was eight years old. I named him Jake, and he taught me a lot about responsibility. When I was 15 years old, my parents allowed me to take in another puppy, a Black Lab named Buddy. To see the difference between Jack and Buddy amazed me. I knew that Buddy would be more energetic, but I never thought that Jack would become exhausted from Buddy. He really couldn't keep up with the puppy, no matter how hard he tried.

At one point, I noticed that Buddy wouldn't leave Jack alone. Even when Jack tried to go off to be alone and rest, Buddy tried to get him to play more or to get his attention in other ways. I quickly realized I needed to train Buddy not to bother Jack when he needed a break. At first, I thought of putting Buddy's e-collar on and shocking him when he would walk near Jack as he tried to rest. But I worried that this could cause Buddy to not want to play with Jack at all. Therefore, I started to train Buddy through a reward system. I didn't give him treats because I didn't want Jack to feel left out with the treats, but I would bring Buddy into another room and play with him or give him a treat without Jack knowing. Slowly, Buddy started to go to Jack when he went to lay down, sniff him, and then walk away.

This experience helped me realize the difference between young and old dogs, and I soon found myself focusing on professional training. Over the past 20 years, I have trained over 200 dogs of all ages. One of the first topics I discuss with all my clients is how the age of the dog matters when it comes to training, but only to a point. You can train older dogs, just like younger dogs, but you have to make sure older dogs get more breaks. They are going to get tired quickly. They are slower, but also determined to make your proud. They need the same consistency, patience, and care with their training that a puppy needs.

Training Your Puppy

Training a puppy is a fun, enjoyable, and stressful experience. While you enjoy seeing them learn and grow, they can also keep you on your toes. They don't have a strong attention span and can easily become distracted. However, they also want to do everything they can to please you, even the more stubborn breeds. Here are a few tips when it comes to training your puppy.

Your Puppy is Not an Infant and Not an Adult

Puppies are at a bit of an in-between age when it comes to the life of dogs. They are still growing and developing on a physical, mental, and emotional level. Yet, they are not infant dogs and can do more than you think by themselves. Because of their age, people struggle to know the best way to train their puppy. I know many trainers who have helped people through the difficulties of training a puppy.

Your Puppy's Developmental Stages

There are several developmental stages that your puppy is going to go through before they reach adulthood. You want to understand these stages so you can get the best out of them when it comes to training.

At four weeks, you should get your puppy socializing with other dogs. If you are a breeder or your dog had puppies, you shouldn't give them away until they are a little over eight weeks old. There is a lot of development that goes on within this time and allowing them to play with their brothers, sisters, and parents will help them grow mentally, physically, and emotionally.

Starting at five weeks old, they should interact more with humans. Of course, if you have children they are going to want to play with the puppies immediately. Try to hold off on letting them around the puppies too much until about week five.

At five weeks old, they will also start to become little investigators. They will get into everything they can and become interested in their environment. Let the roam around as much as possible, but also keep your eye on them. Puppies at this age can fit into the smallest places, causing them to get stuck.

Puppies need the first few weeks of their life to develop their dog-to-dog training skills. This is something you should not interrupt. By week eight, they will have developed more of their primary dog skills and might be ready for a new home. If you bring a puppy home around week eight, continue to focus on their social skills. Don't worry about full-on training yet and do not prepare an e-collar for your dog. Any dog should be at least six months old before you start thinking about an e-collar

By ten weeks, you want to make sure your puppy has had plenty of socializing opportunities and explored their surroundings. If they haven't they will start to become easily afraid of the unfamiliar. This is an anxiety that the dog will carry with them throughout their life, making them more difficult to train.

A Puppy's Fears

Before you start training, you have to understand your puppy's personality, including their fears. Most puppies are going to have some type of fears, especially in a new home. First, if you notice your puppy acting startled when they come across something or a person they don't recognize, this is natural. They will recover easily and try to see what this new stimuli is all about. They will continue to be curious and eventually grow out of the startling phase. However, all dogs will naturally be a little hesitant when it comes to new stimuli.

The trouble with puppies and their fears is if they don't show signs of recovery after a few minutes. There are many warning signs that will alert you to their fear:

- Trembling
- Scanning the room
- Lack bladder or bowel control
- They want to withdraw from the environment
- Whining
- They are trying to avoid what is scaring them
- Excessive panting
- Refusing to eat
- Vomiting
- Diarrhea
- Salivating

If your puppy starts showing some of these signs, of course they won't be potty trained yet, around five weeks old, they may have inherited fearful tendencies from a parent. At three months old, your dog will start to become anxious if you don't do what you can to help them through their intense fear. Once a dog becomes anxious, they will remain this way for the rest of their life.

Don't Start House Training Until Eight Weeks Old

While you shouldn't worry about an e-collar at this point, you can start basic training with your dog at eight weeks old. One of the first steps you will do is train them on their main spot. This is the area where they will eat, sleep, and spend the majority of their time. For example, if they will be an inside dog, they will remain indoors. If they are outside dogs, they go outside the majority of the time. It is essential that you teach them their potty spot at this age.

Training a Puppy Takes a Certain Mindset

Justin opened his own dog training business 25 years ago. The mission of his business is not just to help people learn how to train their dogs with an e-collar, but to overcome the challenges of training. "A lot of people tend to give up because they feel some forms of training are a lost cause. If they can't give their puppy to stop jumping on people, they become more relaxed about it because it's not worth the struggle. I have had people tell me they have other things to worry about than getting their dog to listen to them about everything. This is the wrong mindset to have when it comes to training. I always tell people who say something like this that they need to develop a more compassionate and tougher mindset. They need the compassion so their dog still feels comfortable around them. Compassion can also help them understand their dog needs consistency in training. A tougher mindset is more for them than their dog. This mindset doesn't allow you to give up. You want to remain strong because you want to do the best for your dog."

Training Your Older Dog

While most training tips are pretty general between puppies and older dogs, there are a few factors that are more important for older dogs.

Book 1 - E-Collar Training Step-By-Step

You Can Train Them for a Longer Period of Time

Older dogs have a longer attention span than puppies. While you should start with shorter training sessions, such as 10 to 20 minutes twice a day, you can increase this amount of time little by little. For example, if you start with 10 minutes, two weeks in you can increase to 12 minutes. Always watch your dog and notice if they start to get distracted or too tired. You don't want to force them to go longer than they can handle.

Always remember, each breed is different. Some breeds are going to naturally have longer attention spans than other breeds. For example, Border Collies, Labradors, and German Shepherds have some of the longest attention spans because they have high levels of concentration.

You Need More Patience

You always need patience when training your furry friend, but you need more patience with older dogs. You can teach an old dog new tricks, but it is going to take longer than it will for a puppy. Puppies naturally catch onto new tricks quicker because of their age and curiosity level. Older dogs are not as curious, and like people, tend to slow down.

It's natural to feel like giving up on your training sometimes, you might even feel this way with a puppy. The key is to take a break if you need to—your dog might need a break too. Take a few deep breaths and think about all the process you and your dog have made. Sometimes people look so far into how much work they need to do that they forget about their progress. Don't let this happen to you as it will make you feel like training isn't working.

Socialize Your Dog

Just like puppies need socializing, adult dogs need to socialize just as well. No matter what age your dog is, they need to run, play, meet with other dogs, and people. Otherwise, dogs can suffer from fear, shyness, and loneliness. They won't know how to interact with people or other dogs. Other than the park or having your dog meet your friends and family, take them on walks with other dogs and dog training classes.

Get to Know Your Dog

If you don't know about your adult dog's past, take them to a dog behaviorist and a trainer. They can help you learn to understand your dog's past a little better by the way they act. Training dogs is always easier when you can understand certain behaviors.

Chapter 5: Let the Training Begin

Tips to Prepare Your Dog for Training

While I've already discussed several tips to prepare your dog for training, I want to bring up a few more. Some of these tips will focus on home training and some on obedience training.

Always Have Your Dog's Attention Prior to Training

E-collar training is not going to work if you don't have your dog's attention. When you first use the e-collar, you need a way to grab your dog's attention when you are going to give them a command. You might do this by tugging on their leash if you are outdoors or saying their name. You might decide to use some kind of motion, such as snapping your fingers twice. Once you can consistently gain your dog's attention, you can start with basic training.

Make Sure You Have a Little Play Time

If your dog is too energetic to focus on training, the whole process is going to fail. Play with your dog before you start training. You can take them for a walk or to the park. Do something that is outdoors as this will calm your dog out more than playing inside. Don't give them too much exercise because you don't want to make them too tired for training. The point is to let them run off enough energy so they feel calm and can focus on their training.

Have Everything You Need with You

You want all your training supplies right next to you when you start training. If you have to head to grab the treat, your dog is going to get distracted and then you need to start over. Place the treats in your pocket or hide them in some way so your dog doesn't notice them and focus on the treats instead of training. If you are going to implement the e-collar, have this available to you, if it's not already on your dog's neck. You can even have a few toys near you to reward your dog by playing with them for a while when they follow through with your command.

Empty Your Dog's Stomach

Don't feed your dog right before you are going to train them or head to a training session. If they need to use the bathroom, they will become distracted or could have an accident. While accidents can still happen, feeding your dog a few hours before you start training will help avoid accidents.

Every Dog Should Know the Basic Commands

There are a lot of tricks that you can teach your dog. You can teach them the basic commands or you can become more advanced. In general, what you decide to teach your dog is up to you and your dog. While every dog, no matter what age, should know the basic commands such as stay, sit and lay down, not every dog will want to learn other tricks. For example, your older dog shouldn't learn how to ride a skateboard. You don't need to train your dogs to jump rope, hide their head, wave, shake hands, or hug. You might feel these are fun tricks, but if your dog isn't into it, they shouldn't be forced.

Basic Commands

Finally, after all the information you've received it is finally time to start looking at some basic training strategies. The training techniques you will learn in this chapter don't all need an e-collar. While it is your choice to use an e-collar to teach your dog to sit or not, some professional trainers advise against it while others say it works great for basic training.

Some of the basic training techniques are best used with a reward system instead of a shock if your dog doesn't listen to your commands. However, if you struggle with training and feel you and your dog will benefit from the e-collar, then it is up to you. All you need to do is make sure your dog can handle the e-collar. For example, if your puppy is only four months old, even the smaller e-collars might be too much, but you can still try. You also need to remember to follow your dog. They will let you know if something is too much for them.

Bed Training

Some people feel that the e-collar should not be used for bed training. Like any training you do with the e-collar, it is generally up to you. If you feel you and your dog are ready for the e-collar, then use it for this type of training. You can also use this training to teach them to go into their create.

Step One: Facing the Bed

When you start this training, you want them to face their bed. Point to their bed and say "bed" and give a tone. Once your dog is all the way in their bed, turn your e-collar off. You should never have the e-collar on when they are sleeping. Always remember to praise your dog when they do something they should.

Your dog is going to be confused about the "bed" command at first; this is common when you just start training them in a new task. Don't spend too much time on forcing them to go to bed and don't stress them out by pushing the stimulation button over and over. You also need to be careful about disciplining your dog if they don't listen because they can associate this with bed. At the same time, if you give up you are telling your dog they have options and that's not building a strong training foundation.

You know your dog the best and you will probably come up with a solution to help your dog understand that the word "bed" means lay down and go to sleep. For example, you might find that putting a treat in their bed helps then understand or using hand gestures. Whatever extra training tool you use, you need to slowly reverse the action because you want them to go to bed when you give the command of "bed."

Step Two: Back Farther Away From the Bed

Once your dog is going to bed easily when you stand right next to his bed, you want to back up a bit. You can go into the next room and repeat the command. By this time, your dog will understand what "bed" and the tone means. They will more than likely head straight to their bed because you asked them to and they want the tone to stop. Dogs are great at figuring out how to get things to stop as they know what is in their best interest. Therefore, your dog is going to head to his bed because he hears the tone and the only way to get the tone to stop is by laying down in bed.

You can continue to back farther away from the bed as much as you want. For instance, if you allow your dog to roam all over the house, you never really know where you will be when you need to tell them "bed." Dogs tend to remember the scenery they are trained in and will associate this with the command. This does mean that your dog will understand the word "bed" and know what it means when you say it in the living room, but if you have never said it in this room before, they might not go right away.

Home Base and Perimeter Training

Every dog owner has the fear of their dog running off. Some people fear this so much that they don't let their dog outside unless they are with them and their dog is on a leash. Other people are a little more relaxed and will simply hope their dog stays within the yard.

One of the first basic training strategies that you need to teach your dog is their home base, sometimes called perimeter training. This is when you tell a dog where they sleep, where they can go in your house, where they can't go, where they will go potty, eat, etc. Dogs are not a companion you can walk around the house and say, "You go potty here, you go to bed here, you can't go in this room" and expect them to know. They have to be trained, and it will take time.

This training won't keep them at home. Your dog is going to run if they want and can run. The main point of this training is to help them

Book 1 - E-Collar Training Step-By-Step

understand what their role is in the home. It also establishes the owner/dog relationship.

Step One: What Are Your Dog's Boundaries?

To teach your dog where they can go and can't, you need to make sure you set up boundaries before training and stick to them. For example, your dog can go in the living room, but not on the furniture. They can't go into the kitchen, but can stay in the porch, walk-in coat closet, and entry way. You also decide your dog can go to the basement, but not into bedrooms.

You want to do the same thing with your yard. If you live in town, but don't have a fence, you will want to think about how to keep your dog within your yard. This might be difficult with a leash, but it is possible to do your best in training your dog not to leave your yard.

Step Two: Establish Home Base with Your Dog

Before you use the e-collar for home base training, you need to make sure your dog understands the type of training you will work on. You can do this by verbally telling them and spend a few days or a couple of weeks allowing your dog to become comfortable in the home, at least in the areas they are allowed in.

If you have used the e-collar before, you will understand that you want to get your dog comfortable before using. If you haven't used e-collar training previously, follow the steps in Chapter 2 when getting your dog used to home base training.

Step Three: Mark the Boundaries with Cones, Flags, or Anything Noticeable to a Dog

Your dog isn't going to understand the boundaries unless they are clearly marked. If you don't have a fence, sidewalk, or anything else to mark your dog's boundaries outside, you want to think of using cones or flags. You can then teach them that any place beyond the flag or cone is too far.

Step Four: Walk Your Dog to Their Home Base

No matter where the home base is, take your dog's leash and walk them to their home base. Once they are in the area, give them a treat. You will want to do this a few times throughout a series of days. Don't take 10 to 15 minutes and walk them to and from their home base and this may confuse them or stress them out. Plus, it would give them too many treats and they will become sick. Instead, take a couple of minutes to do this about two to three times a day.

Step Five: Reinforce with Verbal Commands

Some people will combine this step with step four right away. It is up to you to do this. I don't do this because I feel separating them gives your dog more time to understand their training. Remember, you need to have patience when you are training your dog.

You want to keep the commands short. For example, you might say "home" to let them know to go to their home base. If you have their bed at their home base, which is typically the case, you can say "bed." You probably won't want to use the verbal cue for "lay" or "lay down" as this should be saved for when you want to train your dog to lie down and relax around company or in another setting. Other words they will want to recognize when it comes to their home base is "stop" and "come." You will use stop when you don't want them to go any farther and come when you want them to come back.

Some people will also use a hand gesture during this phase. They may point to their dog's home base or move their hand in that direction. The biggest problem with using hand gestures is you can use the same gesture throughout your day and not notice. This will confuse your dog when you are sending them to their home base or trying to tell someone which direction to go.

Step Six: Repetitive Training

In this step, you will focus on perimeter training. This is when you look at the whole area your dog can go to and not just where their bed, food, or water is. When you are focusing on perimeter training, you want to make a distinction between home base and the rest of the boundaries. For instance, you will tell them to go "home," meaning home base, when you want them to go to bed or lay down. You will then walk them around the perimeter. You can allow them to walk freely as this will give you time to tell them "stop" or "come" when they go too far.

This step is going to take a while. You will want to spend at least a week, if not more, focusing on this step. Don't take your dog around too often when you are working on this step. While you want to repeat it for a while, you don't want to spend too much time training your dog where they can go and where they can't go. For a puppy, you should keep all training times to two to three minutes, an adult dog at about five minutes, and a senior dog can handle about 10 minutes.

At this point, you should only use the e-collar if you feel your dog is ready for it. Because this is still a basic training step, though at a deeper level, you don't need to use the e-collar yet.

Step Seven: Introduce the E-Collar

If you haven't done so yet, now is the best time to train your dog on their perimeter training with the e-collar. You want to ensure that they understand their training and are grasping the concept. You also want to make sure that you follow the right steps when introducing the e-collar.

Step Eight: Go Beyond the Perimeter

This next step is debatable for many dog owners as they feel it is a trick more than a test. In this step, you want to test your dog to make sure they do not follow you beyond the boundaries. Why many dog owners have problems with this step is because you need to ask your dog to "come" to you when you are outside of their boundaries.

The point of this step is to make sure they understand the importance of remaining in their boundaries. Another reason is because it helps set up the owner/dog relationship where you give them rules and they follow. When you use the e-collar, you will give them a warning and shock them if they pass beyond their boundaries. Of course, you can always use your verbal cues if you don't feel right testing your dog with the e-collar. Other people won't use the e-collar until they have reached this stage and their dog stays within the boundaries. It's always important to remember that dog training is a little flexible when it comes to you and your dog. You need to figure out what works for you and not what other people say.

Step Nine: Reinforcement Phase

In this step, you will allow your dog to freely move around the perimeter and remain hidden from their sight. However, you want to see where your dog is going. The reason why you want to watch is your dog is so you can give them a warning and shock if they cross the boundary. They won't know it is you and will think it is because they are going farther than the flags or cones allow them to.

You can also use this phase to test what your dog will do if something distracts them. For example, if they are outside, will they run toward a passing car? Of course, you will use your training method to bring them back into their perimeter. If they do become distracted, you know that you have more training to go before you can officially wean them off the e-collar for this training.

Sit Training

Whenever you start a new training technique, you want them to be in their home base area.

Step One: Say the Command

Use your attention grabbing technique to get your dog's attention. Say "sit" once with any hand gesture you are going to use to motion them to sit. This gesture needs to be different than the one you will use to tell them to lay down. When you get them to sit, reward them.

Step Two: Add in the E-collar

When you are training your dog, you can start using the e-collar right away, like we did with bed training, or you can wait until they understand the command. In this training technique, I am adding the e-collar in a bit later to show you how it works and why. You want your dog to understand their new command. You can make them feel stressed if you use the e-collar and they don't understand what they are supposed to do. This is going to get training off to a bad start and make you and your dog become too dependent on the e-collar.

When you add in the e-collar in the middle of training, you want the e-collar to replace your old techniques, such as a hand gesture or a treat. Your dog won't understand the change immediately, so you will want to use both techniques at the same time while slowly decreasing your old technique. For example, the first couple of commands with the e-collar, you will use both techniques. Then, when you feel your dog understands the tone or vibration of their e-collar with the command, you will stop using the other technique. If your dog responds, continue using only the e-collar. If your dog doesn't respond like they were, continue using your old technique at least two more times before repeating this process.

Once you say "sit" send the tone to the e-collar to get your dog associated with the tone (or vibration) and command. Remember to reward them if they sit. Your dog is going to make a mistake a few times and not sit when they are told to. This is normal for dogs, especially puppies, because they are easily distracted and still learning. Even if your dog doesn't listen to the command right away, when they do, follow it with positive reinforcement. Dogs thrive on positive reinforcement, and it will help them learn the command and want to perform.

Step Three: Practice

Just like you did with bed training, practice telling your dog to "sit" in various locations. You can even take them outside of the house to practice as you will have moments of telling your dog to sit wherever you go, such as a friend's house. The key when you practice is to not focus on the task too long and to remain consistent in your methods as much as possible. For instance, there might be times where you don't have the e-collar on and you need to tell your dog to sit.

Lay Down Training

Step One: Home Base

You need to make sure that your dog is in their home base area before you start the training. This is a consistent step whenever you start a new training method as it helps them learn. Plus, they are most comfortable in their home base environment. When your dog follows you into their home base environment, you will reward them. This is always an important step when it comes to your dog successfully completing any step in training.

Step Two: Say the Command

Tell your dog to "lay down." You don't want them to go to bed and lay down because this is going to confuse your dog with the "bed" and "lay down" command. You want to make sure that these two commands are separate and your dog understands this.

When you tell your dog to "lay down" you want to make sure you have their attention and you use a hand gesture or another type of technique they are used to in training. While you can start with the e-collar right way, like we did with the bed training, you can also wait to include the e-collar in the next step.

Step Three: Add in the E-Collar

This step follows the same format that telling your dog to sit does. You will add in the e-collar and turn it on right before you give the dog your command. Remember to reinforce the command with your previous gestures so they understand that the stimulation they feel is telling them to lie down. Again, even if your dog makes a few mistakes along the way, ensure that you always give them positive reinforcement when they follow the command.

Step Four: Practice

Once you are just using the e-collar to reinforce the command, start to practice in various areas around your home, outside, and even at a friend's house. Just as you did with teaching your dog to sit, you will use this command in various locations.

Another factor to remember when teaching your dog to lay down is that they will often follow commands and then get up back and want to play, especially puppies. You may have noticed this when you taught your dog to sit. It's important, especially with lay down, that the dog remains in that position for a few seconds, at least. You might want to train them to stay that way for a few minutes, but you always need to gradually train your dog. At the same time, you don't want to force your dog to lie down for too long, no matter where you are. This can cause them to become stressed, especially if they are energetic dogs. You should always think about your dog's personality when you are training them. This will help you ensure that you and your dog have the best training experience possible.

Come Training

Many people feel it is also important that you train your dog to "come" after you have trained them to "sit." This is because it will give them a sense of freedom from their last command. While some dogs might sit and then get up right away, especially if they notice you have a treat or toy in your hand, other dogs will sit for a period of time or until you ask move or train them to come to you.

Step One: Home Base

You always want to start training your dog at their home base no matter what you are trying to teach them. Most trainers feel that starting at the dog's home base is essential for this training.

Before you start training, choose a spot to stand within their home base. This spot is the area they will come to. If you move around too much during the first few times of the training, you will confuse your dog. They won't understand exactly where they are supposed to go when you say "come." You will focus on moving around later in the training.

Step Two: Say the Command

Following the same method you did with "sit" and "lay down," you will ask your dog to "come" while using a hand gesture or another method of training method they are used to. If you're comfortable starting training with the e-collar, it's fine to include it in this step.

There is a chance that your dog is going to come to you before you even give the command. They don't understand what is going on, so they are going to question why you are standing away from them. This is especially true for a puppy. Bring them back toward their bed or a bit away from you and try again.

When your dog performs the command, give them positive reinforcement. This doesn't always have to be a treat. Dogs love any type of positive reinforcement, including special attention and play time.

If you haven't used the e-collar to help reinforce this training yet, you should do so once your dog has performed the action a few times.

Step Three: Follow-up with an Additional Command

This command can easily make your dog wander, causing them to become confused. They may not understand when the command is done and continue to follow you. Because you always want to make sure your dog understands that the command is over, you can follow-up with another command. You can use any other trick that your dog knows, but most people use the "sit" command.

People will often use "sit" "lay down" and "come" training together. For example, you might ask your dog to "sit" before requesting them to "lay down." This is helpful for a dog because it gives them one step at a time. Some trainers will command their dog to do all three in a row. Of course, you want to do this slowly and make sure you are telling them on type of training at the time. For instance you will tell them to "sit" and then let them sit for a few seconds before commanding them to "lay down." Once they are down for a few seconds, you then command them to "come."

If you do this as practice, you want to make sure you don't place stress on your dog by performing the commands over and over within a small amount of time. For example, you don't want to spend a half hour repeatedly telling them to "sit," "lay down," and "come." You want to stick to a session of two to three minutes, no more than five minutes, for puppies and between 10 to 20 minutes for older dogs.

Stay Training

Step One: Home Base

Just like previous training, you want to start at your dog's home base. If they aren't at their home base, ask them to come to the home base. Make sure you give them positive reinforcement for following your command.

Step Two: Say the Command

You will start to teach your dog to stay with a hand gesture and your voice. Using the e-collar right away doesn't always work for this command at first because your dog isn't going to understand that they need to stay in that spot. Furthermore, many dogs will move around when they feel the tone, vibration, or shock from the e-collar. This will only cause problems within your training. You can include the e-collar after your dog understand what they are supposed to do when you say "stay."

Step Three: Reinforce with Another Command

"Stay" is a great command that is used after another command, such as sit or lay down. This is why most trainers believe you should teach your dog to lie down, sit, or come before you teach them to stay. For example, you command your dog to "sit." Once they perform this command, you tell them to "stay." This lets your dog know that they are not supposed to get up until they receive another command from you or they get tired of staying in the spot and want to move around. Always remember that you dog, especially puppies, need to move around regularly. They won't have the patience to stay in one spot for a long time. They need to get up and move.

Get Down Training

This training is a bit different to do because your dog needs to jump on something or someone before you can train them to get down. You can't really predict where this training will take place, so you don't have to bring them into their home base. But, you need to be prepared to start this training at any time.

Some trainers will cause the dog to jump up on something so they can start the training. You can do this if you want to train your dog in their home base, but they can become confused. For example, if they see you place the treats on the table and you allow them to jump up on the table, but then instruct them to get down they are going to receive mixed signals from you.

Step One: Say the Command

After you catch your dog jumping up on something or someone, command them to "get down." Use a hand signal or some other training technique to help them understand the command. You shouldn't use the e-collar at first for this command. They will need to understand what get down means before you use stimulation from the e-collar.

Book 1 - E-Collar Training Step-By-Step

Step Two: Reinforce the Command with the E-Collar

Once your dog successfully follows the "get down" command a couple of times, add in the e-collar to reinforce the training. For instance, you tell your dog to "get down" but they don't listen, so you use the e-collar stimulation to reinforce the training. Like the previous training techniques, you will back away from the hand motion or other type of training reinforcement you used once you start using the e-collar.

Step Three: Everyone Should Positively Reinforce Your Dog

You know how important positive reinforcement is when your dog listens to a command, but the person they jumped on may not. Ask the person to give your dog some positive reinforcement for getting down when you commanded them to. You should explain this to everyone that your dog jumps on.

Chapter 6: Training Strategies, Levels, and Your Dog

By now, you know that your dog's breed and personality sets the tone for training, even with the e-collar. There are several breeds that are difficult to train, such as the Chow Chow, Akita, Siberian Husky, and Chinese Shar-Pei. You can train any breed of dog, but there are some that will give you a harder time than others. It's not because the dogs don't listen, it's typically because they have a hard time socializing or have a stubborn personality.

No matter what breed of dog you have or how old they are, you should always look into obedience school. These schools not only train your dog, but the trainers also help you with any problems you have or understanding training in general. It's another sense of support, especially when you and your furry friend are struggling with training.

Training Levels

Professional trainers of talk about how there are different levels of training. Depending on what trainer you talk to will depend on how many levels they focus on, but in general there are five. These levels might change depending on what school you go to. For example, parts of level one might make their way to level two. Some schools will often combine the first two levels because they are the easiest. This allows the dog trainers to focus more on the two harder levels. While you and your dog will go through these levels in obedience school, you can incorporate them at home as well.

Level One: Foundation Training

The basic training we went through in the previous chapter focused on tricks your dog will learn in level one. This level is meant to teach the dog how to follow commands give them the basics to become a well-mannered companion. As the easiest level, most dog breeds catch on to this information and start to show signs of learning after the first class. Most dog trainers will not focus on the e-collar during level one. However, they may help you learn how to use it or talk about e-collar training at this level. Some schools will have special classes for e-collars at level one. You can always talk to your dog trainer if you are interested in bringing in the e-collar for training at this level.

Level Two: Skill Building Training

If the obedience school separates level one and level two, this level will build off the skills your dog already learned in level one. For example, the only trick you taught your dog in level one was to sit. This is because level two focuses on the exercises you will give your before they start training. A dog trainer in level two may focus on:

- Teaching your dog to "lay down"
- Teaching your dog to "come"
- Teaching your dog to sit whenever you stop walking

Level Three: Reinforce Reliability and Behaviors

By level three, your dog is ready for the more complex tricks. This is usually the level where e-collars are introduced and you can start to bring your dog's e-collar. Most obedience schools will allow you to enroll your dog in special classes, such as therapy dogs, mental exercises, and dog sports.

You and your dog might go through real-life situations where you will need to use basic commands. For example, your dog trainer has you and someone else act like you are walking your dogs down the sidewalk when one of them starts growling at the other. The trainer will then talk to you about how to handle this situation.

Level Four: Advanced Skills

You should understand that your dog doesn't have to take part in all of the levels. In a typical obedience school, you need to enroll your dog in each level. The only factor is you can't enroll your dog into level two without level one. You do have the follow the rules and prerequisites set up by the school.

It's also a good idea to make sure that you celebrate the accomplishments you and your dog have together throughout the training journey. Some people will celebrate with their dog, maybe by getting a special treat after each level. Your dog may not understand what the celebration is all about, but when you are excited your dog is bound to be excited. Furthermore, making it to level four is definitely something for you and your dog to celebrate!

Level four is the time when you start to show off all of the tricks your dog knows. Throughout this level, your dog will refine their skills, so they can amaze people even more with all their knowledge and tricks. At the same time, your dog will learn a few harder tricks or focus on sitting or lying down for a longer period of time. In this level, your dog may learn:

- Waiting at the door. This doesn't necessarily mean your door at home. They will learn to wait at any door. For example, if you head to the store and you can't bring your dog inside, you can leave them outside to wait. Of course, you always want to consider your dog's safety in this case. If you live in a small and rural area, it might be fine, especially if you know and trust everyone. You can also leave your dog in the care of another person you trust while you run into the store. However, if you live in a larger city, it is best not to leave your dog unattended at all. Many dogs are stolen because they are left unattended in cars or by a store.
- Learning to ignore distractions and follow your commands.
- Walking with you while their leash is loose and ignoring any distractions, such as another dog or people.

Level Five: Expert Training Skills

For most obedience school, this is often the top level. Once you reach this level, your dog has an amazing ability to control their behaviors–most of the time. Remember, your dog is never going to be perfect at training. They will make a mistake from time to time, such as becoming distracted and not listening to your commands. When this happens, it is essential that you use your e-collar as it will motion the dog that this behavior is not acceptable. You should never use the e-collar to discipline your dog because they are not listening. As stated previously, your dog always needs to be exhibiting the unwanted behavior when you use the e-collar. Some of the behaviors your dog will work on during level five are:

- Listening to other people when told to go into a different position, such as when your dog is at the veterinarian's office.
- Staying in the "lay down" or "sit" for a longer period of time
- Sitting until you allow them to move
- Learning when they need to back up so they are not in the way

Practice Real Life Training

Professional trainers will talk to you about practicing real-life training moments before you are surprisingly put on the spot. This will allow you to think through the process you and your dog needs to go through to get the best outcome. When you practice training techniques this way, you can prepare yourself and imagine how you will handle certain situations. For example, if you bring your dog into a store (always make sure the store allows dogs) how will you handle your dog jumping on the shelf and knocking items down? Here are some examples of ways you can practice real-life training. These types of training experiences will work for any breed and dog of any age.

Practice Sitting Politely

One of the first tricks your dog will learn is to sit. One of the easiest training exercises you can bring wherever you go is teaching your dog to sit and wait for your next motion. This is going to be easier for some dogs than others. For example, if you have just started training your puppy, chances are they are not going to sit for a long period. They will sit and then get up a few seconds later. An older dog, especially a senior dog, will sit easier for a period, but it might be a good idea to motion him to lay down eventually. You always want to keep your dog's health in mind when you are training.

Whenever the opportunity arises to get your dog to practice sitting nicely– take it! It's a great way to get your dog used to sitting in various settings. Over time, you will notice him sitting for longer periods of time. This is a great perk about training consistently, your dog is going to adjust and they will find ways to keep themselves occupied when they need to sit or lay down because there is company over. But, no matter how patient your dog becomes, it's important to watch them and notice when they need to get up and move around. If your dog is well trained, such as they graduated from level five, they usually won't get up unless you motion them to. Don't let your dog sit there for too long that it becomes a bother for them.

Case Example: The Mailman and Alice

Alice is a young Miniature Schnauzer who loves to meet the mailman. Typically, this wouldn't be a problem, but Alice doesn't like the mailman to leave. She will try to bite his pants to get him to stay. Because Alice doesn't let go easily, she has started to ruin the mailman's uniform and he becomes a bit behind on his route. While he hasn't complained to Alice's owner, they don't want to cause any problems for the mailman. Therefore Alice's owner contacts a dog trainer they know.

The trainer told them to use Alice's best training skills to keep her from the mailman. "You don't want her to get too close, especially to his clothes. The first trick I would try is getting her to stay sitting when the mailman approaches. Also, keep the e-collar on for when she won't listen, but you need to be cautious about using the e-collar. You don't want her to become afraid of people, and this can happen if you send her the tone when she is heading to him. The best time to use the collar is when she doesn't listen to letting go of his clothing."

The next day, Alice and her owner waited for the mailman. The owner couldn't help but chuckle at her little furry friend because Alice stood so proud as she waited. Soon, she could notice him and became excited. "Sit" Alice's owner told her whenever she stood up. She would listen to the command, but once she became excited again, she would stand back up. By the time the mailman walked into the fenced yard, Alice couldn't contain her excitement anymore. She ran up to the mailman and followed him all the way to the house, which is their routine.

The problem doesn't begin until the mailman gets closer to the fence to leave. Right when Alice's owner noticed her dog about to go for his clothing, they snapped their fingers twice, catching Alice's attention. The owner then said "sit." Alice sat and looked at the mailman as he walked closer to the fence. Alice looked back to her owner, who didn't say or do anything. By the time Alice turned around, the mailman shut the fence door. He waved bye to Alice and carried on with his route.

Over the course of a couple weeks, Alice learned that she could follow the mailman to the house, but when he got close to the fence after dropping off the mail, she had to sit and allow him to leave. Because Alice is well-trained and her owners are consistent, the e-collar never had to be used for this situation.

Use "Come" When You Get a Chance

You don't need to be in a certain spot to tell your dog to "come." Once they are trained in this trick, you can use it when you want your dog to come to you. For example, if you want them to eat their evening meal, you will get their attention by calling their name and saying "come." The key is you need to listen to make sure they heard you. If they are playing in the next room with the kids, they might not hear you call them. Don't assume that you need to use the e-collar immediately. Take the remote and check to see if your dog can hear you call them. If you open the door to a lot of noise and they are playing, it's safe to assume they didn't hear you. Grab their attention and say "come." If they don't follow you because they are too distracted, then you will use the e-collar.

Take Your Dog for Car Rides

Most dogs love to ride in cars, so getting them into the car will not be a problem. If your dog is a little older, you may need to help them into the vehicle. Before you tell your dog to get into the car, you need to think of where you want your dog to stay. You also want to think of what you are going to do if you are driving and your dog gets in your way or a smaller dog tries to jump out of the window. Typically, these instances are rare and you can take precautions so they don't happen. For example, you will have your dog sit in the back seat until you know how they are going to react in the car. Another tip is not to leave your window low so that your dog can jump out.

Taking your dog on a car ride when you need to run an errand is a great way to practice "stay" with your dog. Of course, you need to make sure that your dog is safe and they won't get too hot in the car. Don't stay in the store long as your dog can become upset if they are not used to being alone in the car for an extended period of time. Remember to reward your dog with positive reinforcement when you return to the vehicle. If you have a larger dog, you can bring your dog out to play for a bit. They can stretch their legs or run around and catch a frisbee. It's always a great idea to bring toys with you just in case you have the opportunity to play with your dog.

There are a lot of drive thru places, such as banks, that will give your furry friend a treat and say hi. Even if it is through the window, your dog is interested in the new person and can't wait to get the treat they see!

Car rides are a great opportunity for your dog to learn more social skills. Other than the drive thru, you can take time to stop at a store that allows dogs and let them roam. You need to follow store policy, such as keeping them on a leash, but they will meet other people and possibly other dogs.

Book 1 - E-Collar Training Step-By-Step

Case Example: Minnie and the Store

Minnie is a nine-month-old Poodle. Her companion, Paisley, adopted her from an animal shelter about two months ago. For help training Minnie, Paisley started taking her to obedience classes. Today, Minnie is in level three and struggling, so Paisley decided to take her dog out to a local pet store. She knows that Minnie suffers from a little social anxiety and wants her to become more at ease meeting new people and dogs. Paisley and Minnie's trainer talked about this situation, so Paisley feels prepared.

At first, everything went well. Paisley and Minnie walked up and down the dog isles where Minnie got to pick out a new toy. Paisley is now sitting on a bench where she is practicing "sit" and "stay" with Minnie when an energetic Golden Retriever comes walking up the aisle. At first, Minnie didn't seem too worried but showed interest in the dog. Minnie stood and started wagging her tail. Immediately, Paisley responded, "Minnie" to get her dog to look at her. Once she had Minnie's attention, Paisley told Minnie to "sit." Minnie obeyed, but still paid attention to the dog.

Looking at a dog book, Paisley's eyes scanned Minnie from time to time. She wasn't paying attention to the Golden Retriever until they became interested in Minnie. A young boy struggled to hold on to the leash of the Golden Retriever when the dog took off toward Minnie, causing the boy to let go. The boy started running and yelling at his dog, but Paisley remained calm. She got down to the level of the dogs and carefully watched Minnie and the other dog's reaction to each other. Both of them sniffed the other before they took a step back and looked at each other. Paisley become happy to notice that her dog hadn't moved, except for standing up.

Once the boy caught up to his dog, he grabbed the leash and apologized. Paisley told the little boy it is fine and talked about how the dogs are greeting each other. A few seconds later, the boy's mother came and the family left. Paisley looked at Minnie and praised her for being a good girl with the other dog.

A minute later, Minnie noticed another dog and started to show signs of distress. Paisley looked around and noticed a German Shepherd. This dog sat nicely by its owner and wasn't causing any trouble. Paisley started to care for her dog, trying to make Minnie feel better when the dog stood and started walking closer with its owner. Immediately, Minnie started barking and growling, causing the other dog to become defensive. Paisley started to practice the tricks that Minnie's trainer taught her, such as "no bark," but it didn't work. Minnie remained in defense mode.

Paisley started to feel a little panic, but remembered the trainer told her that Minnie will pick up on the panic, so Paisley remained calm. She then stepped in front of Minnie to create a block against the other dog. Once she noticed Minnie quiet down, Paisley turned around and said, "Minnie" and when her dog looked up, said, "walk" to motion Minnie to walk. Without a second thought about the other dog, Minnie turned and started walking toward the door.

The technique Paisley used with Minnie is known as "creating space" or "blocking the other dog." While this doesn't always work as some dogs will continue to look beyond their owner for the other dog, it worked for Minnie because they had trained for it in obedience school.

Aggressive Dogs

One of the most common problems dog owners have is meeting or noticing their dog becomes aggressive. It is important to note that your dog doesn't have to show signs of aggression prior to becoming aggressive. There are some situations, such as the above example with Minnie, when is it going to happen for various reasons. You may never understand why, but it is important to have a plan in place so you know how to handle these situations.

Paisley knew a bit about Minnie's history. She knew that Minnie lived with a larger dog that was mean to her at times. Therefore, Minnie would naturally become anxious or even aggressive when a larger dog approached her. So, before Paisley decided to take Minnie out to stores or around town where they could practice training in real life situations, she talked to Minnie's trainer, who helped Paisley come up with a few tricks to distract Minnie from larger dogs. Last on the list, was to send stimulation to Minnie through the e-collar. Because Minnie listened to Paisley when she said "go," Minnie did not receive the tone or shock.

Other than creating space, there are several other ways to handle aggressive dogs.

Keep Any Greetings Short

Some dogs don't mind running into another dog because they are curious animals. They want to smell the dog and say "hi," but then they want to continue on with their walk or don't care for the dog or the other owner. When this happens, your dog might become aggressive, just as Minnie did.

At the first sign you see of any type of aggressiveness, such as barking or growling, you need to take your dog out of the situation. One of the biggest mistakes people make is telling their dog "no" or to get angry. This is only going to cause the dog to become more aggressive because they feed off your emotions. At the same time, they might be warning you and they feel like you are ignoring this warning. It's always important to remember that dogs are protective. If they feel something isn't right they are going to take action. There comes a time when you need to listen to your dog and allow them to say "The conversation is over, let's move along."

In general, the best way to avoid any chances of this occurring is to keep greetings short. Even if you meet someone you know, if your dog is unfamiliar with the dog, person or seems uneasy, it is time to move on. Tell your friend you will give them a call and continue walking with your dog.

Another reason to keep greeting short is because your dog can become obsessed with the other dog. While this doesn't always pose a problem with aggression, it can for the other dog. If you do strike up a conversation, move your dog every few seconds and acknowledge them. This is also a good trick to follow if you meet a new dog and someone you don't know. Your dog can become stubborn and not want to continue on the walk because they want to say "hey" to the other dog. If it's okay with the other dog owner, let the dogs meet, but keep your dog moving at the same time. Slowly walk away from the dog and call your dog every few seconds to get their attention.

Notice How the Other Dog Is Acting

Even if you and your dog notice another dog approaching at the same time, you still have room to react to avoid a confrontation. Not only should you notice how your dog is reacting, but also the other dog. If you see any type of signs of aggressiveness from either dog, it's time to move your dog in another direction. You may have to do this by distracting your dog or creating a block from the other dog, so they decide to go across the street. Of course, the first method to try is simply walking your dog in another direction. Typically, dogs are very quick to follow their owners. If you do this before your dog becomes too interested in the other dog, you won't have a problem. One of the biggest keys to avoiding any aggressiveness is to stay one step ahead by noticing the signs. This is usually the best way to avoid any problems.

Avoid Other Dogs

Another way to handle meeting other dogs and dog owners is simply to avoid them. If you notice them, your dog is going to become more interested. If you don't pay attention to them, your dog is going to learn to just keep walking. While they might still look and try to get a sniff of the other dog, it is easier to tell them to come without much of a struggle.

Intermediate Level Tricks for Dogs

It is essential that all dogs receive basic training, discussed in the previous chapter. Now, I want to take training to another level and look at some of the tricks your dog can learn at the intermediate level. As long as your dog is healthy and interested in learning, they can perform these tricks.

However, none of them are necessary. They are tricks that you can use as conversation starters or to surprise people. They are tricks that will allow your dog to show off.

Fetch

Teaching most dog breeds to fetch is pretty easy, it is almost like some of them have this trick in their body chemistry. But, there are other dogs that will struggle learning how to fetch. Not everyone feels the e-collar is necessary for fetching unless there is a possibility of your dog running away or causing trouble with people or other dogs.

There are three main skills your dog needs to know when it comes to learning how to play fetch:

1. Get it
2. Bring it
3. Drop it

Some people like to call the third skill "give it." The trouble with this is if your dog is playing with someone who isn't used to them, your dog could accidently bite the person's hand. For example, your nephews are visiting and want to play fetch with your dog. You agree and everyone heads outside. You've trained your dog to give you the ball, but you have to grab the ball out of their mouth for them to understand. You have never had a problem with your dog accidentally biting you or scratching with their teeth. Therefore, you aren't worried about anything happening to your nephews. In fact, you don't even think about the possibility.

As everyone is playing, you see your nephew hold on the ball in your dog's mouth. You notice your dog acting normal, but suddenly your nephew backs away and starts crying. You run to him, asking what is wrong. He shows you that one of your dog's teeth cut his finger.

You know this is an accident on your dog's part, so you don't discipline your dog. In fact, they look concerned about the situation. You take your nephew in, get the wound cleaned up well, and bandage it up. You know you have to watch it just in case there are signs of infection. You take a moment to think about how this can happen again as the neighborhood kids love to play fetch with your dog. It's at that moment you decide to teach your dog to "drop it" instead of "give it."

Talking to your dog's trainer, they tell you the best way to do this is by walking through the whole training process with them. Your trainer stated, "You don't want to just change part of the game without walking through the whole game. While your dog will have no trouble with the first two steps, you will need to take your time changing "give it" to "drop it."

One key tip for teaching your dog to play fetch is to have two toys that are exactly the same. You will play fetch with one while you hide the other one in your back pocket or somewhere out of their vision. You want them to focus on one toy at the time.

Step One: Tease Your Dog with the Toy

Whenever you want to play fetch with your dog, you need to get them interested in the toy. This is often caused "teasing." You can do this in many ways, as I am sure you have seen other people tease their dog before throwing a toy. You can talk to your dog to get them interested or act like you are throwing it in different directions before you really throw it.

Depending on your dog's age will depend on how excited they get for the toy and game. Older dogs might not be interested in playing fetch. However, younger dogs are going to love this game. If your dog isn't interested, try a different toy. If they are still not interested, they might be tired or want to do something else. Follow your dog's cues to notice if they are interested in something else. If a dog wants to play, they will let you know. If they are tired and want a break, they will inform you of this also.

Step Two: Toss the Toy

Once you get your dog's attention with the toy and they are ready to have it, toss it a few feet. Don't toss it very far at first as you still need to teach them the rest of the steps. When your dog picks up the toy and holds it in their mouth, praise them. You can run up to them and tell them what a good dog they are or use your regular praising method.

Step Three: Bring It

It is now time to teach your dog to bring the toy back to you. This can be one of the hardest parts because if a dog is not used to fetch, they will run off with the toy. Because of this, you want to make sure your dog understand where they can run off to and where they can't. If your dog goes beyond home base and you have their e-collar on, give them the tone to come back without saying or doing anything. They won't associate the tone or shock to their toy or play time, they will contribute it to going beyond their perimeter. Once they come back, praise them and continue to train them on bringing the toy back to you.

The easiest way to train your dog to bring the toy is to say your dog's name followed by "bring it." For example, "Princess, bring it" and when your dog takes two steps toward you, praise them. Go to them, grab the toy, and play a game of tug with them for a few seconds. This is when you gently try to take the toy out of your dog's mouth. The key is to let your dog win, meaning you let go of the toy while it is still in their mouth.

Step Four: Give It

At this point, you will give your dog the next cue to give the toy back to you. Take out the identical toy and tease them with it. This will get them to drop the toy in their mouth and focus on the toy in your hand.

Throw the toy a bit father this time, as you want to pick up the toy your dog dropped and place it behind you without them knowing you have it. Again, when your dog picks up the identical toy and walks toward you, praise them.

Step Five: Use the Word "Fetch"

You will repeat the first four steps by using the key words "get it," "bring it," and "drop it." After your dog has a handle on playing fetch, you will then start using the word "fetch."

When you throw the toy after teaching her, say "fetch, get it." Once your dog has the toy in their mouth, say "bring it," and when they bring it to you, say "drop it."

Once you start using the word fetch, you can slowly increase your distance. If you are at home, remember not to throw beyond your dog's perimeter. If they get distracted or go beyond their limits, use the e-collar to get them back into the perimeter.

Spin and Twist

When you teach a dog "spin and twist" you are teaching them 360 degree turns. This trick isn't typically taught to senior dogs as they don't have good balance. It is a great trick to teach puppies and some older dogs. Again, you know your dog the best. If you feel that this trick isn't for your dog, then you can skip to the next trick.

There are two main skills your dog will need to learn to accomplish this trick:

1. When your hand moves in the counterclockwise motion and you say "twist," they perform this action.
2. When your hand moves in a clockwise motion and you say "spin," they perform this action.

Step One: Halfway Spin

Start by having your dog stand or sit in front of you. With a treat between two fingers in your right hand, let the dog see the treat. Lure your dog in a clockwise half circle. Visually mark the point your dog reached and bring the treat back to you. Drop the treat by your right foot. This should get the dog to face you again. If they become distracted before or after the treat gain their attention. If this doesn't work, use the stimulation on their e-collar to gain their attention. Once they are calm and sitting or standing next to you again, move on to step two.

Step Two: Halfway Twist

Follow the directions from step one, but use your left hand instead of your right. Have your dog turn halfway counterclockwise and then drop the treat next to your left foot. Make sure you have their attention before moving on.

Step Three: Add the Verbal Cues

Once your dog has caught on to your luring, add in the verbal cues of "twist" when your dog moves counterclockwise or left, and "spin" for clockwise or right. Always remember to alternate between right and left.

Step Four: Start Phasing Out the Treats

Start with your right hand. Hold your hand up like you have a treat and continue to lure your dog half way. When your dog is looking at you, bring their attention to your left hand. Make sure they notice the treat so they don't become distracted by looking for the invisible treat. Once your dog makes the halfway turn, drop the treat.

Step Four: Complete the Circle

Keep using the treats in your left hand, but none in your right. Lure your dog through the whole way starting with a spin and then going for a twist. Remember to use your hand gestures and say the word as you are using the whole circle. Don't do this too much as your dog can become dizzy. Once they have follow command a few times, drop the treat and praise them.

Step Five: Add Speed and Mix Directions

Once your dog has a good handle on step four, you can start moving them a little faster and mixing up the direction. This means that you can have your dog twist twice, spin once, twist once, and then spin twice. At this point, you can keep the treat for when they are done spinning and twisting.

Paul Davis

Chapter 7: Advanced Training with E-Collars

Now that you have the basic training and a couple of intermediate level tricks to teach your dog, let's focus on the more advanced training. When you get to this point with your e-collar, you will find that you don't need to use them too often. People will often send their dog the stimulation because they become distracted, aren't listening, or they leave their perimeter. But, because dogs are curious animals and easily distracted, it is always best to keep the e-collar on them when they are in the middle of training.

Will I Stop Training?

One of the biggest questions people ask if they will ever stop training their dog. My honest answer to this is it is really up to you. However, you should always practice the training your dog knows with them. While they will remember to "sit" when you tell them to, even if it's been a few days, if you become sloppy with training then they will follow this path. This means they will focus more on distractions than what you are saying. This will cause you to use the e-collar often. Plus, you will notice all the work you and your dog put into training is disappearing. So, while I do not advise that you stop training, it is always up to you.

Another reason you don't want to stop training is you can avoid having to reinforce any training. If you slack on training or stop telling them to sit, lay down, or stay they will start to ignore the commands. When this happens you will start to have trouble with your dog and become frustrated. You will tell them commands, but they will ignore you. This is when you need to reinforce all the training, starting with the basic tricks. To do this, you will want to start over.

Readjusting the E-Collar for a Growing Dog

Another factor to remember is to watch the e-collar. If your dog is growing, this means that collar has to grow with your dog. While you don't need to go through all the steps to get your dog adjusted to the e-collar, you will want to place the dummy collar on them when you need to clean the collar, charge it, or fix the strap so the e-collar will fit them correctly. Of course, if you don't always have the e-collar on your dog, you can do this when they are sleeping or taking a break.

Agility Training

To focus on agility training, you need a dog that is easy to train. Breeds that strive on obedience are the best dogs for this type of training. Dogs that are stubborn can learn agility training, but frustration can easily set in. Many people who have harder to train dogs give up when it comes to this type of training. It takes a lot of dedication and hard work from you and your dog, especially if they are a stubborn breed, but it is always worth it in the end.

Because of the in-depth process of agility training, I will not focus on how to teach your dog certain tricks. Most people will go to an obedience school to get the best agility training because of its advanced nature. Instead, I am going to give you tips to help you decide if agility training is right for your furry companion.

First, agility training is teaching your dog to run through obstacles. Dogs that perform on dog shows are taught in agility training. Some of the obstacles dogs usually learn are:

- The pause table
- Dog walk
- Tunnel
- Weave poles

Book 1 - E-Collar Training Step-By-Step

Basic Training Before the E-Collar

Like with most training, the majority of people like to get past the basic training before they start using the e-collar. Instead, you will focus on a regular form of training you use, such as hand gestures, verbal cues, or a clicker. Clickers often work great because it is easy to gradually switch from a clicker to an e-collar.

Agility Training Is Great for Your Dog's Health

One of the biggest reasons people get their dogs into agility training is because it helps them stay healthy. Because they are so active, they burn off a lot of energy and they stay in shape. One of the most important factors is that you have to make sure your dog eats right or they will struggle when it comes to agility training.

There Are Risks

Unfortunately, agility training is a type of training that involves a lot of risks. Dogs need to pass a check-up by a veterinarian to make sure they are in good physical health for the tricks they will learn. This is an important part that you don't skip. For various reasons, dogs can have brittle bones and if they fall or trip over something while training, they could easily hurt themselves.

You also want to think about the heat. Your dog is going to run and stay highly active throughout their training, which tends to last longer than typical training. They are pushed by the trainers and will require more water, especially if it's hot outside. They will also need more breaks than normal if this is the case.

Using the E-Collar

When your dog becomes comfortable with the basic level of agility training, it's time to bring in the e-collar. Using the e-collar for agility training is different from other forms of training. For example, when you use the e-collar for home base training, you send the tone and shock to your dog when they leave their home base and don't come back when called. For agility training, you want them to keep running. You want them to run, jump, and follow the obstacles as much as possible. Instead of correcting your dog's behavior with the e-collar, you are redirecting their behavior.

When you redirect your dog, you are focusing on their agility route. They are following the rules and are running with you. For instance, Tammy's dog, Barnie, is learning his first agility route. As a puppy, Barnie is easily distracted and interested in almost anything he sees. Therefore, he tends to struggle following Tammy as she runs his route with him. When Tammy started using the e-collar, she decided should would call Barnie and ask him to "come." If he didn't listen, she would send him the signal from the e-collar. This immediately makes Barnie turn back to Tammy and they continue on his route. Within a week, Barnie is running the route beside Tammy without too much distraction. Within a couple of weeks, he is running the route by himself. Of course, Tammy still keeps a close eye on Barnie and will send him signals when he becomes distracted. Once Barnie reaches this point, Tammy and his trainer start adding new skills to his routine.

Adding New Skills

As the story about Barnie shows, you want to make sure your dog understands the basis of their route and know they need to keep going, ignoring distraction, before you add more skills. When you do get to the point of adding more, make sure that you only add one skill at the time. Once your dog incorporates this skill into their routine, then you can add another skill.

It is important that you continue to use the e-collar throughout this type of training. It is often known as the "silent enforcer" because you don't need to say anything. The second your dog goes off path, you send them the signals and watch them head back into their routine.

Roxie Jumped Over a Bar

Roxie's owner, Dustin, wanted to know if his dog would be good for agility training. After talking to Roxie's trainer, he received some tips and a guide to teach Roxie how to jump over a bar. The trainer stated if Roxie can accomplish this task, she is one step closer to agility training.

When it comes to agility competitions, dogs will jump as high as 24 inches. However a few have jumped higher. Dustin's goal isn't to get Roxie to jump high, so he decides to set the bar at its lowest setting, which is something the trainer advised him to do.

Step One: Walk Over the Bar

To get Roxie used to going over the bar, because all she wanted to do was go around it, Dustin got her leash and walked over the jump bar with her. At first, Roxie was a little hesitant to walk over the bar, but once she saw Dustin go over it, she followed.

Dustin repeated this action a couple of times and then started to say the word "over" each time they walked over the bar. Once they were over, Dustin gave Roxie positive reinforcement. Before he knew it, Roxie was interested in going over the bar again, specifically for the positive reinforcement.

Step Two: Roxie Walks Over the Bar Alone

Next, Dustin set Roxie in front of the bar and told her to "sit" and "stay." He then placed the bar on the next highest setting and said, "Roxie, come." The minute Roxie started to go over the bar, Dustin said "over." Dustin would repeat this action a couple of times before moving the bar up a notch.

Step Three: Roxie Jumps Over the Bar

Getting Roxie to jump over the bar wasn't the easiest part of the training, but Dustin understood this. Once the bar got to Roxie's elbow, she decided to dive underneath the bar. Dustin rewarded Roxie for her efforts and had her try again. This time, Dustin said "over" right as Roxie got to the bar. Roxie stopped, looked at the bar, and tried to jump over. Unfortunately, she knocked the bar down. Again, Dustin rewarded Roxie for her efforts.

It is at this point that Dustin starts to incorporate the e-collar. Instead of giving a visual cue, Dustin is going to say the word "over" when Roxie is supposed to jump. While she might not make it every time, if she refuses to jump, Dustin is going to give her a signal. Dustin's goal is to eventually stop saying the word "over" and simply give a signal to get Roxie to jump on time.

Step Four: Using the E-Collar

Dustin decided to help Roxie a little more by getting down to her level. Standing on his knees near the bar, he told Roxie to "come." Once she got to the bar, he told Roxie "over" and gave a signal. Roxie walked up to the bar, but simply looked at Dustin. He then repeated the signal. Right as Roxie jumped up, Dustin said "over" and gave another signal. He found Roxie on the other side. She had finally jumped over the bar without tripping or knocking it over. Dustin rewarded Roxie, and they continued this process a few more times without lifting the bar.

Boing

If you want to train your dog to jump over the bar but notice they struggle with their jump, then you can try a trick known as boing, which is when your dog jumps in the air on cue. When people use the e-collar for this type of training, they will usually switch to the e-collar instead of using a verbal cue. However, it is recommended that you start the training by using the verbal cue of "jump" or "boing" and not the e-collar.

Step One: Lure Your Dog to Jump

To lure your dog to jump, it's best to have a treat in your hand. You will crouch down a little and then spring up onto your toes. When you jump, you want to make sure that your arm is up as this is more likely to make your dog jump up because they want the treat.

Step Two: Use Verbal Cues

Once your dog starts to jump with you, add in the verbal cue "boing." Continue to jump as you were until you feel that your dog can jump up to your verbal cue.

Step Three: Use the E-Collar

When you start using the e-collar, you will still want to use the verbal cue as you send your dog a signal. Once you have repeated this action a few times, you can start to get your dog to jump higher. You will continue to use treats to accomplish this mission. For instance, you can first hold the treat between your fingers and raise your hand to where you want your dog to jump. Once they get high enough, you can hold the treat in the palm of your hand. By this point, you should have to use any verbal cues. Instead, you will focus on the e-collar's signal.

Hunting

Similar to agility training, you are going to use the e-collar for hunting once your dog understands the basics. It is also meant to redirect their behavior and not correct it. One of the basic tips for hunting with an e-collar is to get one specifically designed for hunting. While they are more expensive, you can reach your dog up to one mile, which is great when you want to call your dog back to you.

Teach Your Dog the Route

Before you take your dog out hunting, make sure that you planned out a route. You can start by taking your dog on a walk, preferably with a leash unless they are great at staying by your side and not running off. While you can use the e-collar, because you are training them for hunting, most people like to use the e-collar later in their training.

You want your dog to become used to the route. Take your time and let them smell and mark their territory. Don't hurry them along the route because they can easily forget it. You should take your dog on the route a few times before going onto the next step.

Make Sure to Train Your Dog to Return Home

Taking your dog hunting can be a scary time for a dog owner, especially at first. There is always a possibility your dog can get a bit lost–or you lost from your dog as they have a great smelling nose to find their way back to you or their home. However, you should always make sure that your dog understands how to return home. Before you start this process in your hunting training, you will want to make sure you and your dog understand home base training.

Another tip when it comes to training your dog to return home is to have a special signal. For example, you might give them two signals if you want them to return home instead of one.

Repetition Is Key

Like with all other types of training, repetition is key when it comes to hunting training. You always want to make sure you take time to show your dog what they need to do on their hunting adventure. Many owners will spend time every year retraining their dog for hunting, especially if they only go deer or duck hunting. This type of repetition is mainly for safety and to ensure your dog remembers what they are supposed to do.

Chapter 8: Common Mistakes

By now you understand that the e-collar is one of the biggest training tools for dogs. While there is controversy, if you use the e-collar correctly, your dog will thrive. Unfortunately, no matter how much you learn about e-collar training, mistakes can happen.

Lack of Consistency in Training

One of the most common mistakes is a lack of consistency in training. Because I have talked about consistency through this book, I won't spend too much time on it here. However, it is important to discuss as a mistake.

No matter where you are, if your dog does something that they aren't supposed to, you need to inform them of this correctly and immediately. Dogs are not going to understand what they did wrong 15 minutes ago. They live in the moment and need to be corrected right away. If you wait a few minutes, you are going to confuse your dog and they will associate the shock to whatever they were doing at the time. This can cause a lot of issues if you're not careful.

Training Your Dog for Too Long

It is important that you limit the amount of time you train your dog, especially in one setting. Younger dogs shouldn't be trained in one setting for more than five minutes while older dogs can generally last about ten minutes. If you struggle with time, you can set a timer on your phone. This will alert you to when it is time to give your dog a break.

When you don't pay attention to the time, you can overwork your dog. This can cause them to become stressed, tired, and sick over time. Plus, if it is hot you need to make sure that your dog gets plenty of breaks to cool down and water. They will need more water than normal if they are training outside on a hot day.

Dog Owners Don't Send a Signal Immediately

One of the most common mistakes happens when the transmitter isn't close to you and your dog misbehaves. For instance, you are in the kitchen when you notice your dog in the living room chewing on the couch cushion. You quickly scan the room for your remote and notice it is on the table in the living room. As you walk into the living room to grab it, your dog turns their attention onto a chew toy on the floor. Once you grab the remote, you push the button to give your dog a warning and shock. Unfortunately, your dog is no longer chewing on the cushion, so they associate the shock to their own toy.

If you don't keep your remote with you at all times, there will be moments you don't get to train your dog when they take part in unwanted behavior. This is something you want to avoid doing. Even if you are in your home, have the remote clipped to your pants or in your pocket so you can quickly grab it when you need to. This will not only make sure you can catch your dog when you need to, but will make training more consistent.

You Wait Too Long to Start Training

It is sometimes hard to know when the perfect time to start training is. There are general rules of thumb, but this doesn't mean it's right for every dog. One key is you want to start training your dog as soon as you bring them home. Even if they are potty trained and only eight weeks old, you need to make sure they start learning home base, where to sleep, where to eat, and the rules of the house right away. Many people don't think of these factors as training, but they are. Whenever you are teaching your dog something new, it is a form of training.

People Become Codependent on the E-Collar

The key to using the e-collar as a training tool is to get your dog to stop the unwanted behavior. Once the behavior as ceased, you need to wean your dog from the e-collar. If you continue to use the e-collar or you use it for all forms of training, you will find yourself become dependent on the e-collar. When this happens, your dog is also going to become dependent on the e-collar. Dogs who are dependent on the e-collar need to feel the warning vibration, tone, or shock to know they are doing something they shouldn't. If they don't feel one of these, they will continue to take part of that behavior.

You Don't Give Your Dog Enough Training Time

When you start training, you need to understand that you will train your dog often and you won't stop. Even when your dog has learned the basics, you will continue to train them. One of the biggest mistakes people make is they train their dogs the basic commands, when they want them to sit, lay down or get down, and then don't focus on any actual training time. This doesn't give your dog enough time to really learn the foundations of training.

Always spend time every day training your dog. Enroll them in an obedience class and get to know a professional dog trainer for extra help in case you ever need it.

The Dog Has Not Received Any Type of Prior Training

The e-collar is not meant to start training your dog the basics. It is meant to help your dog understand that certain issues are not appropriate, after they have received other types of training. For example, you will not use the e-collar on your dog when you are potty training them. First, most dogs are potty trained within a couple months of age, meaning they are too young for the e-collar. Second, using the e-collar when you catch your dog going to the bathroom can cause them to feel like that behavior is wrong. They won't associate the shock to the fact they didn't use the right are to go; they will associate it with their actions. This can make dogs feel that they are doing something wrong whenever they need to go potty.

Most people who use the e-collar right away, meaning without any prior training, are people who believe the e-collar is used as punishment or they don't understand the e-collar. You should never believe that the e-collar should be used as a form of punishment. Most e-collar companies advise against this and all dog trainers do.

You Use Harsh Discipline

It's a given that you will need to discipline your dog from time to time. For example, you might send them into a special kennel for a "time out" when they scratch the wall. However, most trainers state the best ways dogs learn is through training and positive reinforcement. While you may not completely believe this, you should never use any type of harsh discipline, such as hitting, staring down, grabbing them by the neck, jerking their leash, or yelling. Using this type of discipline can cause your dog to become aggressive and fearful. You can also harm your dog through some of these actions.

If you need help in understanding how to gain better control over your dog's behavior, the best place to go is obedience school. They will help you learn the ropes of training and make sure that your dog understands what behaviors are acceptable and which ones aren't.

People Don't Understand How the E-Collar Works

One of the most common mistakes people follow is getting the e-collar and immediately putting it on their dog and using it. Sometimes people will try it out, even if the dog isn't taking part in unwanted behavior, to see how their dog reacts. This is a huge mistake and something you should never do. You must allow your dog to get use to the e-collar before you use it. Furthermore, you have to wait until they are doing something you want to change before you give them a warning and then shock. If you are going to use the e-collar to train your dog, always use it correctly.

Book 1 - E-Collar Training Step-By-Step

Chapter 9: Frequently Asked Questions and Answers

There are many common questions people ask about training their dog. Trainers find this part of their job enjoyable because they want to help people and dogs have the best training experience possible.

Question #1: Do Different Breeds of Dogs Learn Differently?

A lot of people wonder if they have a certain breed of dog, if they need to train them in a certain way. The straight answer to this question is no. All dogs, no matter what breed or age, learn the same. They can learn the same tricks with the same steps, tips, and strategies. As long as you are consistent and follow the steps within this guide, you can train any breed with an e-collar successfully.

Questions #2: Do I Wean My Dog Off the E-Collar?

Yes, you shouldn't continue to depend on the e-collar if you don't need to. If you feel your dog is successfully trained in the task, slowly wean them off the e-collar. You can do this by placing a dummy e-collar on your dog when you would have given them the e-collar. Remember, your dog shouldn't realize that the e-collar is what gave him the shocks. They should believe it was their behavior. As long as you trained your dog successfully with an e-collar, weaning is a breeze.

You can place the e-collar aside until you need to train your dog again or you find yourself bringing home another new family members. Remember, when you are not going to use the e-collar for a period of time, you want to make sure it won't go off randomly, meaning turn your e-collar off and put it away.

Question #3: How Do I Know the E-Collar Isn't Harming My Dog?

The only way you will know is by watching and understanding your dog's reactions to the e-collar. If they whine, become frightened, cry, or show any signs of distress when you shock your dog, the stimulation levels are too high. Simply lower the level to the lowest setting and adjust as you need to. The dog should only slightly move its head when it feels the shock.

Most dogs respond to the lowest level of stimulation. While this does give them a little discomfort, it does not hurt them. Most trainers will tell you that a little discomfort is fine when it comes to keeping your dog safe and healthy.

At the same time, most trainers will tell you if you are afraid to use the e-collar, you shouldn't use it. As stated before, try the e-collar on yourself before you place it on the dog. This might ease your mind about how it will hurt your dog.

Question #4: How Do I Know When to Start Using the E-collar?

The simple answer is when you feel the time is right. Most people say that dogs should be at least six months old before they receive the e-collar because of the weight and size of the collar. The key is to do your research and make sure that you spend enough time going over your preparations for e-collar training. For example, you know your house rules, you know how you want to train your dog, and you are looking for a dog trainer to help you with the process.

Of course, you don't need to have a professional dog trainer on your side. However, it is always a great idea to get your furry friend enrolled in a training class for extra support and social time.

Question #5: I Have Seen Other Dogs React Negatively to the Shock from the E-Collar. How Do I Know My Dog Won't?

This is a great question. The truth is, a dog can react negatively to the e-collar for many reasons. First, they are not trained properly with the e-collar. For instance, the stimulation level may be too high or they receive a shock for every little thing they do wrong. You should never use the e-collar in this way. You only want to use the e-collar for certain training procedures.

Second, the dog may be more sensitive than other dogs. Just like humans, each dog has their own personality and there are several dogs that are highly sensitive. This will cause a dog to feel the shocks strongly, stronger than other dogs. This will also cause them to internalize the shock more than other dogs, causing them to become sad or even depressed. In general, if you have a highly sensitive dog, you might not need to use the e-collar. Highly sensitive dogs tend to follow instructions through a person's tone of voice a lot better than anything else.

Conclusion

If your head seems like it is overloaded with information, this is typical when it comes to dog training, especially if you are beginning this journey. Fortunately, this book is here for you whenever you need it as you can simply download the book and have it in your phone or on your device for whenever you need it.

Even if you feel you might need to reread some information before you start training your dog, especially when it comes to the training steps, I know you have a better idea about e-collar training and how you can incorporate the e-collar into various tricks.

One great fact about this book is it not only explains what e-collar training is, but you also receive a number of tricks to teach your dog. This book didn't simply focus on the basics, it also looked at more advanced training techniques.

Some of the most important takeaways from this book are:

- Making sure you dog becomes used to the e-collar before you bring it into training. You never want to put the collar on and start giving your dog signals. This is going to cause your dog stress as they will be confused about what is going on. Furthermore, they will know that you are doing this to them and the trust between you and your dog will start to fade.
- You have to be consistent as a dog trainer. It doesn't matter if you are in your home or at a friend's house, when your dog does something they shouldn't, you need to be quick and correct their behavior. Remember, dogs only focus on what is going on at the moment. If you give them signals from the e-collar a few seconds after the unwanted behavior, the dog is going to associate the

signals to what they are doing at that moment. Always notice what your dog is doing when you are in the middle of training, and be ready to act to help them learn that some behavior is not acceptable.

- E-collars will not harm your dog unless you have the collar on too tight or the stimulation level too high. Always make sure you can fit two fingers between your dog's neck and the collar. Start the stimulation level at the lowest setting and wait until you need to use it on your dog to correct their behavior. If they turn their head or react in a slight way, they noticed the stimulation. If they didn't respond at all, it is time to adjust the level. Remember to only move the stimulation level up one. If your dog whines or whimpers when you give a signal, the level is too high and you need to bring it back down.

- Make sure that you don't buy the wrong collar for your dog. Some e-collars are not meant for small dogs. Dogs that are under a few months old shouldn't be given an e-collar. While most people feel the general rule is about six months old, you can start your dog on the e-collar earlier. It really depends on the amount of training they have. It is important to note that there are tons of great e-collars available to you. There are so many that I couldn't include them all in this book. Do your research to make sure you really do pick the best e-collar for your dog.

- Always find a way to gain your dog's attention before you start training them. Even if you've trained your dog for two years, you always want to grab their attention. Dogs are naturally easily distracted, and it isn't fair to them to send them signals when you didn't let them know that you wanted them to follow your command. Use their name or find another method, such as snapping your fingers twice, to gain their attention.

- Always reward your dog with positive reinforcement when they are training. Even if they make a mistake, you want to reward their efforts. You don't always need to give them a treat. Taking time to play with them and talking to them is a great way to give your dog positive reinforcement.

Book 1 - E-Collar Training Step-By-Step

- While dogs can receive training at any age, you always want to keep your dog's age in mind because at the end of the day, it does matter. Puppies are going to be more energetic, which can make you feel more frustrated with training. Older dogs are going to be slower, which means you need more patience. Another factor about older dogs is while they can train for longer periods of time, they will tire out sooner. Never push your dog to finish training or train longer when they are showing your signs that they are tired.
- Start training your dog as soon as you bring them home. Even if they are only 10 weeks old and are potty trained, you want to focus on the basics, such as where they eat, sleep, and their home base.
- Do whatever you can to make sure your dog eats healthy and gets enough sleep, as much as you can do this for a dog. Eating healthy and getting enough sleep will help ensure that you and your dog have the best training experience possible. Furthermore, this is something to do with your pets whether you train them or not. It is up to us to make sure they get food with the nutrients they need so they can live a long and healthy life.
- Don't become codependent on the e-collar. While you will use it often, most people believe the point of the e-collar is to correct your dog's behavior and then wean them off. If you become too dependent on the e-collar, then your dog will only react when they receive the stimulation. You want your dog to react to your commands and not the stimulation of the e-collar. Always remember the e-collar is more of a reinforcer when it comes to training.
- Dog training takes a lot of patience. You will need to repeat the steps over and over to give your dog the best training experience. Always remember that the number one factor with training is the safety and health of your dog. Once you keep this in mind, you can give yourself a breather when you start to feel frustrated or losing patience. You are trying to help your dog and not harm them in any way. Rushing through training is going to harm them mentally and emotionally. Furthermore, if you have the e-collar on too high and find yourself giving them signals often, you could end up

- physically hurting them.
- Follow the directions on how to put on the e-collar to ensure it is in the right location. If you place the e-collar on the base of your dog's neck, it will move up the neck when your dog is playing, This can cause them problems and they won't feel the signals.
- Enroll your dog in an obedience class. This will not only help your dog when it comes to training, but it will help you as well. You will meet a great dog trainer who can help you through some of the toughest times and meet like-minded individuals. Everyone you come in contact with in obedience class will become a part of your journey. Furthermore, your dog will get some social time, which is important. If you only have one dog, your dog can become lonely and feel isolated if you don't allow them to play with other dogs and meet other people. Obedience training is one of the best ways to do this.

Above all, you need to enjoy your time with your dog. Do what you can to make dog training fun. Of course, it will be more work that you can imagine. However, as you work with your dog and listen to them, they will listen to you. If you take the advice given to you about how to train your dog in basic commands as well as more advanced command, you can watch your dog thrive with everything he learns. Through the tips given to you in this book and remembering the key takeaway notes, you will have the best training journey possible with your furry companion. This will give them some of the best memories throughout their life. You will have a great relationship and your dog will spend many years as one happy, obedient, and loving dog.

Book 1 - E-Collar Training Step-By-Step

References

5 Tips for Dog Training Session Prep. (2017). Retrieved 1 October 2019, from https://acmecanine.com/5-tips-dog-training-session-prep/

Accessories & Replacement Parts | E-Collar Technologies. Retrieved 28 September 2019, from https://www.educatorcollars.com/accessories

APDT - "Real Life" Training. Retrieved 2 October 2019, from http://www.trainyourdogmonth.com/tips/reallife.aspx

Arterburn, J., & Benson, K. Dog Shock Collar Myths and Misconceptions. Retrieved 21 September 2019, from https://www.securepets.com/debunkingmyths.html

Barkless Dog Collar | Barkless Pro Anit Bark Collar - E-Collar Technologies. Retrieved 26 September 2019, from https://www.ecollar.com/products/barkless-pro-anti-bark-collar

Bender, A. (2019). Our Top 10 Puppy Training Tips. Retrieved 29 September 2019, from https://www.thesprucepets.com/top-puppy-training-tips-1118511

Bennett, J. 7 Must-Read Tips When Buying a Purebred Dog. Retrieved 26 September 2019, from https://www.rover.com/blog/purebred-dog-tips/

Choosing a Puppy. (2019). Retrieved 17 September 2019, from http://www.apdt.co.uk/dog-owners/choosing-a-puppy

dog agility training for beginners: how to get started - SitStay. Retrieved 2 October 2019, from https://sitstay.com/blogs/good-dog-blog/dog-agility-training-for-beginners

E-Collar for Dogs - Remote Training Collars | E-Collar Technologies. Retrieved 27 September 2019, from https://www.ecollar.com/categories/remote-dog-trainers

Evans, B. (2018). Benefits of using Dog Training Collars that you can't ignore. Retrieved 27 September 2019, from https://petspy.com/blogs/dog-training/benefits-of-using-the-e-collar-training-that-you-can-t-ignore

Finlay, K. 5 Reasons You & Your Dog Should Take An Obedience Class. Retrieved 29 September 2019, from https://iheartdogs.com/5-reasons-you-your-dog-should-take-an-obedience-class/

How to Care for a Formerly Abused Pet. (2016). Retrieved 29 September 2019, from https://www.tasteofthewildpetfood.com/training-behavior/how-to-care-for-a-formerly-abused-pet/

Howell, A. (2019). The E-Collar Dog Training Bible : The All-Inclusive Guide, Including Specific E Collar Training For Golden Retrievers, German Shepherds, Labrador Retrievers, And Beagles. Kindle Edition.

Hunting Dog Collars | E-Collars for Hunting Dogs - E-Collar Technologies. Retrieved 26 September 2019, from https://www.ecollar.com/categories/hunting-dog-trainer

Krohn, L. (2017). Everything you need to know about E Collar Training. Retrieved 2 October 2019, from

Obedience Courses. Retrieved 2 October 2019, from https://www.animalhumanesociety.org/behavior/obedience-courses

Pavia, A. (2017). Adopting A Shelter Dog: 5 Tips For Success. Retrieved 23 September 2019, from https://fearfreehappyhomes.com/adopting-shelter-dog-5-tips-success/

Puppy Training and Socialization Tips for Owners. (2011). Retrieved 29 September 2019, from

Book 1 - E-Collar Training Step-By-Step

https://healthypets.mercola.com/sites/healthypets/archive/2011/09/22/every-puppy-owner-must-know-about-early-training-socialization.aspx

Ray. (2010). History of the Shock Collar. Retrieved 28 September 2019, from http://dogtrainingclub01.blogspot.com/2010/12/history-of-shock-collar_20.html

shibashake. Dog to Dog Aggression – Why and How to Stop It. Retrieved 2 October 2019, from https://shibashake.com/dog/dog-to-dog-aggression

Stregowski, J. (2019). Are You Guilty of These Dog Training Mistakes?. Retrieved 2 October 2019, from https://www.thesprucepets.com/common-dog-training-mistakes-4030442

Things to Consider When Choosing a Dog. (2017). Retrieved 17 September 2019, from https://trupanion.com/blog/2017/03/choosing-dog/

Train your dog: The relevance of consistency! | Tractive. (2018). Retrieved 28 September 2019, from https://tractive.com/blog/en/training-en/consistency-and-rituals-in-dog-training

Transitioning for Canines. Retrieved 29 September 2019, from https://primalpetfoods.com/pages/transitioning-for-canines

Where to get a puppy. Retrieved 28 September 2019, from https://www.humanesociety.org/resources/where-get-puppy

Wildesen, A. The Misunderstood Doberman Pinscher. Retrieved 17 September 2019, from https://thecaninetrainingcenter.com/the-misunderstood-doberman-pinscher/

Wilson, S. (2019). 8 Things You Need To Know Before Buying A Shock Collar | CanineJournal.com. Retrieved 27 September 2019, from https://www.caninejournal.com/shock-collar-for-dogs/

Paul Davis

BOOK 2

TIPS AND TRICKS TO DOG TRAINING

A How-To Set of Tips and Techniques for Different Species of Dogs: Based on Real Experiences and Cases

PAUL DAVIS

Book 2 - Tips And Tricks To Dog Training

Introduction

You are sitting on your couch watching your new puppy run around the house. They only came home a few hours ago and already into everything. As you watch your dog run from one side of the room to the next, you can't help but smile and wonder what you got yourself into. Can you handle a puppy? Of course, you can! You can handle training any dog with the tips and tricks laid out in this book.

Like with every new topic, you have to start with the basics. You will find this information in Chapter 1. You will receive information on what you need to know before training and common training methods. You will gain an idea of if you are interested in the electric collar, relationship-based training, or positive reinforcement training. It is important to note immediately that every type of training focuses on positive reinforcement at the end of the training step or session.

Chapter 2 gives you more information about training and your dog. There are dog breeds that are easier to train and some that are more stubborn. The type of breed you get does matter when it comes to training. For example, if you want to train a dog specifically for hunting, you will want to look into a dog whose main instinct is to hunt. These dogs, such as Labrador Retrievers, are the easiest dogs to train for hunting. While Poodles are great dogs for training, they are not the best dogs to train for hunting. However, they are great dogs for agility training.

Another part of training your dog is to make sure you are caring for your dog correctly. You will want to keep in mind that they need a certain diet, especially if they are agility training. You will also receive tips about keeping your dog's mouth clean and performing weekly health check-ups, which is especially important for your older dogs.

And what about those training challenges? Every dog, no matter how easy they are to train, is going to come with their own training challenges. How do you help your dog overcome these challenges? You will learn about this within the pages of this book.

Chapters 3, 4, 5, and 6, focus on training dogs at different ages. For example, Chapter 3 looks at training puppies and mainly focuses on the basic training, such as sitting, because dogs are only puppies for a short period of time, and they don't receive a lot of training time. Chapter 4 looks at training dogs during their adolescent phase and continues with basic training. Chapter 5 goes into training an adult dog and Chapter 6 looks at training a senior dog. Each of these chapters also hold training tips and other great information to help you and your dog through the training journey.

Chapter 8 is here to help you understand some of the common mistakes dog owners make throughout the training process. These mistakes each have ideas to help you overcome the mistakes. The key to making sure you don't fall victim to some of the common mistakes is to remain mindful of your training. For example, if you are going to use the electric collar to train your dog, you want to have the remote with you whenever the e-collar is on your dog. By remaining mindful, you will realize when you don't have the remote with you and find it before you need it to train your dog not to dig in the garbage.

Now, it's time to take a deep breath, realize you and your dog will make a great team for training, and focus on the tricks and tips within this book. While you do not receive all the tricks that your dog can learn (there are hundreds of tricks) you will get a variety of tricks to keep any level of dog at any age happy and healthy.

Chapter 1: The Basics

Training your dog can make you feel overwhelmed, it doesn't matter if you are focusing on basic training, such as sitting and laying down, or advanced techniques, such as dog sports. There are moments you will become frustrated, no matter how patient and calm you feel. Then, there are those wonderful moments—the times your dog succeeds at a new trick. It's these moments where you feel that all your hard work has paid off.

But why do people focus on the end result when they are training? Professional trainers will tell you that you have to celebrate every moment your dog made an effort with their training, even if they made a mistake.

Celebrating the missteps your dog takes during training won't make them put in less effort. It won't tell them that this moment is "good enough." It will tell them that you are proud of them for trying. You're proud of them for working hard and you want to reward them with positive reinforcement. Doing this will get your dog focused on the task and want to make you happy again.

Celebrating the moments with positive reinforcement, even when a mistake is made, is one of the first points you need to understand when it comes to training.

What to Know Before Training

Celebrating is not the only point you need to understand before you start training. You need to understand that your dog's personality depends on how well they will train. You need to know about training tools, how your dog's age matters, and how your mindset matters.

Put Thought Into the Name

Naturally, you will choose a name that you love, but if you know you'll spend a lot of time training your new family member, it's important to think of a name that will catch your dog's attention. Names with strong consonants that are short work great. They will perk up your dog's ears, making it easier for you to catch their attention. Some great names to consider are Ginger, Jack, and Jasper.

You may think you can only give your little eight-week-old puppy a name and not your older dog you received from the shelter. It's important to know the name of your dog from the shelter tends to be temporary. While this isn't always the case, most shelters don't know the dog's name when they receive them. Call your dog by their shelter name and notice their reaction. If they don't respond, they don't know that you are calling them, and you can rename your dog. Even if you have a dog that responds to a name, you can still change it. This can give you a good start on learning how to train your dog.

Have One Consistent Way to Grab Your Dog's Attention

Dogs are easily distracted by their environment, especially puppies. This causes problems with training because their owners feel they aren't listening to them. In reality, your dog became distracted and didn't know you were talking to them. By setting up a way to grab your dog's attention, such as calling their name, you will train them to look at you.

For example, you notice your dog is in the yard sniffing something on the ground, but they are too close to the road. Because the way you get your dog's attention is to snap your fingers, you head outside and snap your fingers twice. Even though you are several feet from your dog, they are used to this sound and know you are calling to them. They look up and listen to your command to "come."

Consistency Is Key

One of the strongest ways to effectively train your dog is remaining consistent. It is always possible you won't catch your dog in action or be unable to reach your dog when they misbehave, meaning you have to let this one slide as they won't understand what they did wrong. When it comes to these moments, you need to realize they happen. The trick is to not let them happen often. Professional trainers say you have about five seconds to correct your dog's behavior. Once this time passes, you need to let it go and try to catch them in the act the next time.

Patience Is Another Key

Another strong key feature when it comes to training is patience. If you have expectations that your dog will learn to sit by the end of the day, you must lower your expectations. It will take days to weeks to train your dog one trick and you will never stop training.

You will need more patience for a dog you just brought home as it takes them time to adjust to their new environment and for an older dog. Senior dogs are slower than puppies. They aren't going to catch on to new tricks as quickly and can show more stubbornness because they are used to their ways. However, with patience, consistency, and positive reinforcement you will teach your old dog new tricks.

Prepare for Your Dog to Come Home

Just as people prepare to bring their baby home, you want to prepare to bring your dog home. This doesn't mean you need to set up a whole room, but make a spot for their bed, crate, food, water, and some toys. Your dog needs their space just like you need your space. It'll help them feel more at ease and comfortable in their new environment. Plus, they will immediately start to feel that you care.

You hear about dogs sleeping in bed with their companions, but this isn't the idea you want to give a dog you're training. Teach them everyone sleeps in their own space. If possible, give your dog space that isn't around other people or pets. It's common for people with more than one dog to set up their kennels or beds right next to each other. If your dogs laying next to each other brings comfort to them, it's fine to allow it. But, make sure that both your dogs understand they each have a bed and space. There will be times they want to be alone.

Other ways to prepare for your dog to come home:

- **Plan the arrival**: If you work during the day, try to bring your dog home on a Friday night or Saturday morning. Spend the weekend getting to know your dog by taking them around their new environment, observing their behavior, and playing with them. Give them as much love and support as possible because they are frightened and uneasy about their new home. They need to trust you before they can truly enjoy their new surroundings.
- **Gather any supplies you will need:** At first you may think your dog needs a couple of dishes for water and food, a bed, crate, and a few toys. However, there are a lot of other supplies you can bring into your dog's environment immediately to help them prepare for training from the start. For example, a leash, collar, and any other training tools you will choose.
- Also, a couple of old towels or rugs are a good idea for several reasons. It can give your dog's space a more comfortable look and be quickly available in case your dog has any accidents. Even a potty-trained dog can have accidents when they first come to a new house.
- **Think about your other pets:** Animals are sensitive, and they will get jealous of the new family member if you don't give them enough attention. It's common to tell ourselves our other pets "will adjust" and we "don't need to worry." The truth is, you should worry about how all your pets feel. Try to play with all of your pets, make sure you give everyone equal attention, and talk to them when they walk by. Animals love to hear your voice and it makes them feel comfortable.
- **Make sure everyone is healthy:** Even if you are told the shelter gave your dog a check-up, bring them into the veterinarian's office as soon as possible. Not only can you ensure your dog is healthy, but your vet and dog can start their friendship. You should also take your other pets in for a check-up, especially if it's been a few

months or more. It's always a good idea to make sure everyone is healthy for the new arrival. Unless you are going to breed, think spaying or neutering your dog if they aren't already. The veterinarian will guide you through this process.

Don't Push Aside the Crate

It's easy for you to see a crate as a jail cell for a dog, but dogs see a crate as a place to call their own. As long as you don't make your dog stay in the crate most of the day, you will find your dog lounging in their crate at times.

One factor to consider with a crate is how you will use it. If you want to use the crate as a way to punish your dog for bad behavior, you won't want them sleeping in the crate. In this case, you will want to think about getting a bed and a crate or two different types of crates. You don't want to tell your dog to go to bed in the same crate you give them a time out in because they can confuse the two meanings. If your dog feels you are angry with them during the night, they aren't going to sleep well, causing them to become sick or depressed.

Think About Your Training and Discipline

Some people use training as a way to combat disciplining their dog. They believe training their dog to listen to their commands and using a firm voice is all they need. For most dogs, this will work as long as you are training your dog correctly and consistently. Even if this is the way you want to go, it is helpful to look into the best ways to use any form of discipline. You also need to be aware that certain forms of discipline will have serious consequences.

Correctly disciplining your dog is not an easy task. It will take planning and ensuring that every family member is on the same page. This is something you will also need to do with training. To effectively discipline your dog, consider the following tips:

- Your dog shouldn't know that you are disciplining them. If your

- dog realizes it is you, they will misbehave when you aren't around. This is one reason why a lot of people turn to training in order to teach their dog how to behave.
- Your dog lives in the moment. If you discipline your dog a minute after the unwanted behavior, they will associate the discipline with their current action and not their previous action.
- Do not use discipline with an aggressive dog. This can make their aggression worse. If you find yourself struggling with an aggressive dog, it's time to think about obedience classes or a professional trainer who can help you.
- You need to teach your dog a wanted behavior to replace the unwanted behavior. This is another reason people like to use training. However, you should never think of training as a form of discipline.
- Don't severely discipline your dog for a small crime and don't weakly discipline them for a big crime. This is a difficult imaginary line to find, but it is necessary. Think of it this way: If you are too harsh on your dog, you're going to make them frightened and weaken any trust and respect within your relationship. If your discipline is weak, it won't cause the dog to work on changing their behavior.

There are many ways that you should never discipline your dog because it will cause problems in the future.

You should never hit your dog, not even a pat on the nose. Dogs are highly sensitive animals and this action will cause them to trauma, especially if it happens continuously.

Never knee or kick your dog. You can easily hurt them, and it will cause them trauma.

Don't "rub your dog's nose in it." If your dog has an accident, look over your training procedure or take your dog to the veterinarian to ensure it's not a medical issue.

Don't yell at your dog. This is often our automatic reaction when they run into the street or do something that's unsafe. Yelling is similar to hitting or kicking with a dog, as it can easily traumatize them.

Don't jerk the leash when your dog pulls you on a walk. Take time to train your dog so they don't take part in this behavior.

Common Dog Training Methods

There are many methods you can use to train your dog. In fact, you may find yourself overwhelmed by all the methods and struggle finding the best one for you and your dog. You might find yourself using one method, but not liking it after a while because you don't feel like your dog is responding to the method. While it is likely your dog needs a different method, it's important that you analyze your training method before changing on your dog.

There is a lot of disagreement with the various methods I discuss in this section. For example, many people won't use the e-collar because they think the shock will harm their dog. While it can, if you use the e-collar correctly your dog will barely notice the shock and still respond by stopping the behavior.

Positive Reinforcement Training

You always want to give your dog positive reinforcement when they do something well. Therefore, this type of training should be a part of any other form of training. However, it can also stand alone. This type of training has been around for a long period of time, but it's gained popularity over the last few years.

When you focus on positive reinforcement training, you are constantly giving your dog rave reviews when they show good behavior. The main idea behind this training method is that dogs will naturally repeat behavior when they receive positive reinforcement. As long as you remain consistent with the positive reinforcement, your dog will follow the behavior that gives them the attention they desire.

When a dog exhibits unwanted behavior, you simply don't respond. You don't talk to them, you don't pet them, and you don't give them any type of positive reinforcement. This is one of the biggest problems people have when using positive reinforcement. They don't understand how you can ignore your dog's bad behavior and it will go away.

It is important to note that the more stubborn breeds will have a harder time learning through this method because they are more likely to continue their negative behavior. While they love positive reinforcement as much as an easily trainable dog breed, they won't care that you are ignoring certain behavior. Instead, they will see this as an open window to continue the behavior. Dogs that are easily trainable and more sensitive will notice you don't pay attention when they perform certain actions and more than likely stop the behavior or at least decrease it.

Using positive reinforcement training means you have to be on your toes with your dog at all times. You will need to supervise them well to give them positive reinforcement within seconds of their good behavior. If you don't catch them soon enough, they won't understand why they are receiving the extra attention from you and will associate it with the behavior they exhibited at the time.

When you start rewarding behavior for training, you will give them positive reinforcement every time they exhibit the behavior. After they catch on to the training and their behavior becomes more regular and natural, you will start giving them positive reinforcement every other time, then every third time, etc. You want to gradually decrease the positive reinforcement you give with that behavior. This is another piece people struggle with during positive reinforcement training. People want to give their dog attention all the time, but they can't when using this form of training.

E-collar Training

The e-collar is known as the electrical collar and delivers signals, such as a tone, vibration, and shock to your dog's neck. It's important to realize the e-collar is not meant to be a way to discipline your dog. Unfortunately, there are people who use their e-collar for this purpose. They will give their dog a shock every time they do something wrong. The e-collar is meant to be a teaching tool, one that trains your dog to stop the unwanted behavior without realizing it is you giving them the shock. Instead, they believe their behavior caused the shock, making them stop the behavior over a period of time.

There are various types of e-collars and you need to choose the one that is right for your dog. For example, if you are training your dog to hunt, you will purchase a hunting e-collar. For focusing on house training, you will purchase a yard e-collar. You need to look at the basics of the e-collar because each one is a bit different. For instance, a yard e-collar ranges to ½ to ¾ of a mile. This means you can send signals to your dog with the remote until they pass this point. The furthest e-collars tend to go is about a mile. These are typically the e-collars used for hunting.

If you choose to use the e-collar, you want your dog to become used to the device before you start using it. It is advised that you let your dog wear it for a week before you turn it on and start training. You want to train during this time as the e-collar is not necessary for early training. When you start using the e-collar, set it on its lowest stimulation level. If your dog reacts by turning their neck a bit, they can feel the shock and you shouldn't increase the stimulation level. Most dogs will feel the shock at the lowest setting.

Do not use e-collars on dogs younger than five to six months old. While some smaller e-collars will fit on them, most people feel the training is still in its early stages where it is best to use verbal cues and hand gestures to train your dog.

Case Example: Tallie's Garbage Training

Tallie is a Border Collie who loved to go through the garbage inside and outside of the home. After trying to get Tallie to stay out of the garbage for months using their regular training method, they decided to get an e-collar. When the e-collar arrived, Tallie's owners read the instruction manual. They placed the e-collar closer to the top of Tallie's neck so it wouldn't slip when she was playing. After making sure the collar was not too tight by placing two fingers between her neck and the e-collar, the couple allowed Tallie to get used to her e-collar for a week. They kept the e-collar off and never used it to train her to stay out of the garbage. Instead, they continued to use their regular method.

After the week passed, Tallie's owners turned her e-collar on and waited until they found her in the garbage to send any signals. Through their research on e-collars, Tallie's owners understood they had to send the signal at the right moment, or the training would have poor consequences. Therefore, Tallie didn't receive her shock until her nose was in the garbage. Her owners worried if they shocked her walking up to the garbage, she would become afraid of garbage cans and be uncomfortable in certain areas of their home.

When Tallie received the shock, she moved her head to the side a bit, got down from the garbage, and shook her head. At the lowest stimulation setting, Tallie's owners knew that she felt the shock, so they did not need to raise the stimulation level.

Tallie's e-collar stayed on her neck during the day. When she went to bed, her owners placed a dummy collar on her neck to give her the impression it is the same collar. In the morning, Tallie's owners replaced the dummy collar with the e-collar and continued focusing on her garbage training. Within a few days, Tallie started to hesitate when she got close to the garbage. Sometimes she would go the other way and sometimes she jumped into the garbage and started digging until she received a shock. Within a couple of weeks, Tallie stopped getting into the garbage.

Relationship-based Training

Relationship-based training is a new type of training that takes a lot of observation as it focuses on your dog's body language. You train your dog by the way they react, so you can get a better understanding of the way they think and establish stronger communication. Many people refer to this type of training a philosophy in dog training. You can use this type of training for changing old behaviors and forming new behaviors. You do this by reinforcing the behavior you want to see. This type of training works because dogs strive on making sure their owners are happy.

Relationship-based training informs people that every interaction you have with your dog is a teachable moment. You need to become mindful, so you are aware of the way you are acting and treating your dog, other people, other animals, and yourself. Your dog notices everything from your actions to your emotions. Dogs have a very strong sense when it comes to their owner's emotions. They can tell how their owners are feeling and can quickly sense when something is not right.

Here are a few tips for relationship-based training:

- **Place your animal's immediate needs first:** If you find that your dog is acting uneasy, fearful, or looking for a place to hide you need to cease training. It is more important that you try to understand why your dog is reacting this way. They are telling you that something is wrong and one of their immediate needs is not being met. It is important to solve this problem before continuing with training.
- **Learn what motivates your dog:** Observing your dog's behavior and how they react to stimuli will help you realize what motivates your dog. For example, you are training your dog to sit down every time you say the word "sit." Each time your dog sits when given the command, you praise your dog by playing with them and telling them how proud you are and what a good dog they are. Even if they don't understand every word you say, the tone of your voice and your actions inform your dog that you are happy with their behavior. They will remember this and the next time they hear the word "sit" they will follow the action that gave them praise.
- One time, you decide to change your method with training a little. Instead of giving your dog praise, you give them a treat. Every time

you tell them to "lay down" and they listen, you toss a treat at them. Over time, you notice that your dog reacts stronger to the praise than the treat. When praising them, your dog catches on to the training quicker. Therefore, you continue to use praise when training your dog instead of giving them a treat.

- **Learn how to interpret your dog's body language:** Every dog will show signs of how they are feeling, such as sadness, anger, and fear. Unfortunately, there are a lot of myths surrounding the ways to read your dog's body language. Receiving the wrong information will cause you to misinterpret how your dog is feeling, breaking down a trusting and respectful relationship.
- When you decide to train through the relationship-based method, you have to do your research to understand how your dog's breed reacts when they are feeling a certain way. Try to stick to the scientific based information and talk to professionals. You can start with a dog trainer as they can help you or guide you to the right information and people to talk to.
- **Change your dog's environment to prevent their unwanted behavior:** This idea is similar to "child proofing," but you "dog proof" your dog's environment to prevent them from acting a certain way. For example, your dog likes to get into the garbage located in their space, so you take the garbage out. The other garbage bins your dog gets into should have lids and be shut in cupboards or closets, giving your dog limited access to these areas.
- **Supervise your dog carefully:** Don't allow your dog to run around the whole house without supervision, especially if they get into everything or are not potty trained. Keep them with you as much as possible and follow them around the house if they start to wander. You don't have to do this for your dog's whole life. Once they go beyond the basic training level, you will decrease supervision. The point is to keep a close eye on your dog until they have the necessary life skills to know when they need to grab your attention to go outside or to stay out of the garbage.

Like other forms of training, you need to be realistic with training. This will help you overcome any obstacles because you will have lower expectations for how quickly your dog will catch on to training or how often your dog will listen to your commands. No matter how well you train your dog, you will feel there are moments they have selective hearing.

Dominance Training

Dominance training is also known as alpha dog training. This type of training focuses on the instinctual pack mentality of dogs. While your dog isn't around a pack of dogs, they see the family unit as a pack. With this knowledge, the owner can become the alpha, causing the dog to listen to what they say and do. This is another training method where you use your dog's body language to learn how they are feeling.

When you focus on this training, you will want to make sure your dog sees you go first, as the alpha always goes first. For example, you will leave the room first, enter the room first, and receive your food first. You will even walk before them when they are on the leash. If your dog wants to go outside, they have to sit by the door quietly until you let them go outside.

Dominance training is hard for some people because you need to make your dog feel they are underneath you. For instance, when your dog is afraid, one of the best techniques to calm your furry friend is to get down to their level. When you use dominance training, you can't get down to your dog's level, no matter how they feel. You always need to tower over your dog to assert your authority.

Dominance training is starting to decline in the modern world. New studies are coming out to show that dogs don't follow the pack mentality like wolves, making this training ineffective (Clark, n.d.). Professional dog trainers don't usually agree with this training because they feel it is out of date. Most trainers like to follow the scientific studies on how dogs react to training as this gives them better results.

Chapter 2: Training and Your Dog

One of the biggest factors to determine how well training will go with your dog depends on their personality. Your dog's personality depends on the way your treat them and their breed. There are some breeds that are easily trainable and other breeds that harder to train. However, every dog can learn tricks. The key is how you handle the training.

Dog Breeds and Training

It is important to note that all dogs enjoy pleasing their owner and this factor helps when it comes to training. No matter what dog breed you have, you can use this knowledge to help train your dog. If you are looking for an easier time when training, you will want to focus on the easily trainable dogs.

Border Collie

The Border Collie is a smart, energetic, and affectionate dog. They thrive on positive reinforcement, so are a strong candidate for this type of training. They are sensitive dogs and want to do anything they can to make sure their owners are happy. They generally have a height of 18 to 22 inches, so they are great dogs for any type of living. But if you live in a smaller apartment you need to make sure your dog gets enough exercise. You can do this by bringing your dog outside often and taking them to obedience classes.

Border Collies are great dogs for agility training. They love to perform tricks and play in dog sports. They are great show dogs, as long as they are healthy and well cared for. They are in the herding dog family, making them feel they need to herd children in a large group.

Border Collies have more energy than people realize. They are fast runners and quick to catch on to what you are teaching them. They will amaze you with how much they can handle in a day. Many people who train Border Collies say they will finish a task and then look at you as if they want a new task. They don't tire easily, making them one of the best dogs for training.

German Shepherd

The German Shepherd is often used for training for police and military work. They are great dogs to train due to their courageous, confident, and smart nature. Because they are large dogs, reaching anywhere from 22 to 26 inches, they are best in a bigger home. They can adapt to apartment living but will need a lot of time to run around and exercise.

One of the reasons the German Shepherd makes the list of easily trainable dogs is because they are extremely loyal. Once they understand the behavior they are supposed to exhibit, they will do their best to follow this behavior. Furthermore, they can focus on tasks easily and have the ability to learn several tasks in a short period of time.

Unlike some breeds, German Shepherds can focus longer on tasks, allowing you to train them for a longer period of time. Of course, you always need to take your dog's age into consideration. Puppies are going to naturally need more breaks than older dogs, even German Shepherds.

It is important to know when your dog is becoming tired because they won't tell you they need a break. They will keep going until you give them a break. When you feel your German Shepherd needs a break, it is best to stop training and allow them to rest.

German Shepherds have a keen sense of danger and have no trouble putting their life on the line to save the people they love. They don't become scared easily, so they can handle most types of consistent training.

Doberman Pinscher

Doberman Pinschers are loyal, and fearless. They aren't afraid to take on new challenges and are great dogs for a source of protection. Due to their larger size, reaching anywhere from 24 to 28 inches at the shoulder, apartment living isn't the greatest for them. But, with enough time to run and exercise outside, they will adjust.

One cautionary tale is many people believed only experienced trainers should focus on Doberman Pinschers. While they are easy to train, consistency is more important for this breed. They do get distracted easier than other breeds because of their protective nature. They will notice every single noise coming from their environment and keep their eye in this area until they know everything is fine.

Another reason is because Dobermans are known for their aggressive nature. In reality, any breed will become aggressive under certain circumstances. The way you treat your dog will shape their aggressive natures, especially if you get them as a puppy. Dog's learn through how and what they are taught. If you are aggressive with your dog, even in play, they will become more aggressive. This is why it is always important to talk to your dog calmly and never react in an aggressive way toward your dog.

The Pembroke Welsh Corgi

The Pembroke is a smaller dog that can live in any type of environment. They love the apartment life as well as the country life. However, they also need a lot of room to explore or they will become bored, which can lead to depression in dogs.

The Pembroke needs to have a job to complete to stay happy. Even if you aren't training them at the time, it is important you play and interact with them often to give them a sense that they are doing their job. The Pembroke is part of the herding dog family and will try to herd people. However, with the right consistent training you can teach them when and when not to herd.

The Pembroke is a great dog if you are interested in agility training because of their enthusiastic nature. While they are easily trainable, you always need to make sure you have the role of boss. Pembrokes feel they have to be the boss and won't have any trouble trying to take that title away from you.

Norwich Terrier

The Norwich Terrier is a smaller dog that is suitable for any type of living situation. They are great with children and are energetic, even when they feel tired. It is easy to get a Norwich Terrier to jump up and complete a task, but it is also important to not overwork your dog when training. Norwich Terriers will allow you to do this because they strive to make their owners happy. Therefore, it is often up to you to watch for signs to note that your little companion needs a break.

They are intelligent dogs and make for a great breed to train for dog sports. They not only need physical exercise, but also mental exercise. In fact, they won't be truly happy unless you challenge them physically and mentally. Even though the Norwich Terrier is small in size, they are great hunting dogs.

Labrador Retriever

Labrador Retrievers are friendly, outgoing, and active, making them easily trainable. As larger dogs, reaching between 21 to 24 inches at the shoulder, they are more comfortable in a larger space. However, they can adapt to apartment life. Just as any other medium to large sized dog, you need to make sure to take them out for enough exercise. In fact, these dogs love to see how far they can go.

The Labrador Retriever is a great dog for any type of training. They are natural hunters, making them easy to train for hunting. Furthermore, they enjoy taking part in dog sports and shine in agility training. They are sociable dogs and enjoy spending time with humans and other animals.

What Makes a Dog Easily Trainable?

There are dozens of dog breeds that are considered easily trainable. If you want to focus on a dog that is known to train easily, here are some factors to consider:

- **What drives your dog's instincts?** Each breed has their own instinct triggers. Some of these triggers are stronger for your dog than others. When your dog is faced with stronger triggers, they will have a harder time focusing. For example, if a dog is ruled by their nose, they will have a harder time concentrating on a task if they smell something new in their environment. They will feel the need to learn what this new smell is before they can learn how to sit.
- **Is your dog easily distracted?** Puppies are naturally more distracted by features in their environment than older dogs. But there are also many breeds that have to focus on a new sound they heard or what they saw moving.
- **Your dog's personality.** One of the biggest features when looking for an easily trainable dog is to do a little background research on your dog's personality. There are dogs that will cooperate better than other breeds and some that will try to take over control. Of course, the dogs that are known to cooperate will give you an easier training time.

Caring for Your Dog

Most of these tips you will want to follow whether training your dog or not. It is important to make sure your dog is fit, healthy, and happy. When focusing on a healthy and happy dog, there are ten components that you need to follow:

1. Security and safety
2. Nutrition
3. Love
4. Exercise
5. Veterinary care
6. Grooming
7. Comfortable place to sleep

8. Games and toys
9. Family
10. Just leadership

All of the tips outlined below are going to focus on these components so your dog can have the best life and training experience possible.

Choose the Best Food

When a dog is overweight, it affects their health and lifespan. The extra weight on your dog will put pressure on their joints, causing arthritis. Their risk for heart disease, tumors, and skin diseases increase. They will also stop exercising as much because they feel tired and out of breath sooner. This will only increase their chances of gaining weight because they won't run their weight off like they should.

Your veterinarian will talk about the best nutrition for your dog. They may also have food that you can purchase there. It's important to follow the diet that your dog's veterinarian prescribes as many dogs need a certain diet, especially if they are training.

Some of the best dog food brands include Natural Balance Original Ultra Whole Body Health, Horizon Complete, Whole Earth Farms, and Nature's Variety. These foods are generally more expensive than your traditional food, but they give your dog the best nutrients to keep their body healthy and bones strong.

Keep a Clean Living Environment

It's not only important that your environment is clean for the people in the home, but also for your furry family members. Animals will get into anything they can. They don't understand what is bad or good for them. Even if you see your dog sniff something bad for them and walk away doesn't mean they won't try it again. A clean living environment is a basic quality for a healthy life.

Another part to this is ensuring your dog is groomed. There are some breeds who need to be groomed on a weekly to monthly basis and others who can go a few months without seeing the groomer. If you don't want to take your dog to the groomer often, note how much they shed, if they need their fur trimmed, and other grooming factors before bringing your dog home. Give your dog a bath, trim their claws, and brush their coat.

Veterinarian Check-ups Are Important

Don't skip a check-up with your dog's vet! Even if they are healthy and happy, they need their annual check-up. Dogs that are in agility training and older should have a check-up at least twice a year. Once you bring your dog to their veterinarian, they will help keep you on track with any visits for your dog.

Keep Your Dog's Mouth Clean

Many people don't think of their dog's mouth when it comes to the animal's health. But, keeping their mouth clean is essential. Oral problems can cause pain for your pet and make it hard for them to eat or even drink water. Your dog's oral health can lead to kidney disease and increase their chance of death.

While most dog owners don't think it is necessary, it is important to brush your dog's teeth. You can give them a hard treat to help this process, but you should always take the time to brush your dog's teeth at least once a day.

Perform Weekly Health Checks Yourself

Once a week, check your dog's eyes and ears for any type of discharge or redness. Check your dog's skin and coat to ensure there are no signs of scabs, fleas, or ticks. You can always use medication to help prevent your dogs from getting fleas and heartworm. Your dog's veterinarian will help you get the right medication and shampoo for your dog. When your dog eats, watch for any signs that show they aren't eating or drinking like normal. If you notice anything different with your dog's eating or grooming habits, take them in for a check-up.

Book 2 - Tips And Tricks To Dog Training

Tips to Overcome Training Challenges

There are a lot of challenges that happens during training. Most of them you will correct over time through training. For example, you need to call your dog's name a few times before they respond to you. This might happen because they aren't used to their name. It can also happen because they don't want to respond to you. Overcoming this challenge involves a training technique where you play the "name game."

Teaching Your Dog Their Name

There are various ways to teach your dog their name. One of the most common ways is known as the "name game."

1. Start calling your dog's name is a happy and fun tone when they are not looking at you.
2. When your dog turns to you, say their name again and give a verbal cue, such as "good" or "yes" followed by a treat.
3. Repeat this process for two to three minutes a few times a day.
4. Once your dog responds to their name after the first time you say it, stop saying your verbal cue and give them a treat.
5. Slowly decrease how often you give them a treat after saying their name. For instance, every other time, every third time, etc. until you stop giving them a treat.

You can use a different form of positive reinforcement for training. While a lot of people pick a treat, other people worry about their dog eating too many treats. Your dog can get sick if they take in too many treats a day. The best option for treats is to use small ones or break apart a larger treat.

Pulling on Their Leash

Another common trouble when training is your dog pulling on their leash. For example, you are talking to a friend while on a walk with your dog. Instead of your dog sitting there nicely until you decide to move again, your dog starts trying to pull you along with their leash. This is behavior that you need to stop for several reasons.

First, it doesn't show your dog the owner-dog relationship you should have. It places your dog in control and your dog cannot be in control when you train them as the training will not work. Second, your dog can hurt themselves by pulling. Third, your dog can get loose if the collar isn't on correctly.

To stop this behavior there are many forms of training you can use. For instance, you can teach your dog to sit, use a harness as that doesn't allow them to pull as much, or train them that when you stop, they stop. Most people like to teach their dog how to sit, but this doesn't mean that your dog will sit while you are talking. Teaching your dog to sit for a long period of time is a gradual training technique that will take months.

"Sit" is one of the first commands people teach their dog because it helps them in many situations. For instance, you can use it when you are taking a break on a walk or when your dog wants to jump on your company. There are several ways to teach your dog to sit, but these steps are the most common:

1. Get to your dog's level.
2. With a treat in your hand, hold it near their nose.
3. Tell your dog to "sit" and begin moving the treat up slowly. They will follow the treat with their eyes.
4. The more you move the treat, the closer their bottom will get to the floor. If they don't sit all the way down, gently push them down to sit. You need to watch how high you hold the treat because your dog will jump up if it goes too high.
5. Once your dog sits, even with your help, give them the treat and then praise them for sitting.
6. Like with other types of training, you will need to repeat this process regularly throughout the day.

Book 2 - Tips And Tricks To Dog Training

Obedience School

No matter how much training you provide in your home, it is always a great idea to enroll your dog into obedience school. Not only will obedience classes help you train your dog, but you can receive support from like-minded people. You can also discuss any problems you are having with your dog's personal trainer.

When it comes to most obedience training schools, there are about four different levels. As your dog graduates from one level, you can enroll them into the next level. You do not have to enroll your dog in all four levels.

- **Level one, basic skills:** Level one focuses on the foundations of training, such as sitting, laying down, and get down. You will learn about training your dog to stay in their home base and walking your dog with a loose leash.
- **Level two, building the skills:** In this level, you will build on the skills your dog learned in level one. For instance, instead of sitting for a few seconds, your dog will learn to sit for a couple of minutes. They will also learn to automatically sit when you stop walking and coming to you when you call their name.
- **Level three, advanced skills and special classes:** Just like you do in level two, you will continue to build on your dog's previous learned skills. They will learn to sit and lay down for longer periods of time. You can also enroll your dog in special classes, such as agility training, hunting, therapy, and mental games.
- **Level four, expert skills:** In this level, your dog will learn to come when they are distracted. They will also learn how to wait at the door. This isn't just the door to your home; this is any door. For example, if you go to a neighbor's house and they invite you in, your dog will sit outside of their door until you come out. It is always important to keep your dog's safety in mind when leaving them out of your sight. While your dog reached the expert training level, situations can still arise. Most dogs in puppy mills are taken when dogs are left unattended outside.
- Another trick your dog will learn at this level is to back up when

they are in the way. It's common for your dog to walk in front of your feet and nearly trip you, especially at the puppy stage. Dog's don't understand that they can hurt you or you can accidently hurt them when they are in the way. This form of training alerts them to your movements and they will take a few steps back when they are in the way.

Find a Trusted Personal Trainer

Another tip to help you overcome challenges is to find a dog trainer that you can trust. Most dog trainers will take time to privately help you with training or allow you to email or call them when you are in need of assistance. You can ask them questions about how to overcome challenges or why your dog is reacting in a certain way.

Different Ways to Reward Your Dog

You always want to be cautious about how often you give your dog treats. They can get sick, especially if they have a sensitive stomach, from too many treats. Furthermore, if you focus on a lot of training throughout the day, your dog will fill up on treats and won't be hungry for their nutritious meal. Even if you have healthier treats, problems can still arise, such as your dog not wanting to eat their food because they want the treats instead. Therefore, it is best to come up with a different way to reward your dog other than using treats. Here are various forms of positive reinforcement:

Play a Game for a Few Minutes

Instead of a treat, you can play your dog's favorite game for a few seconds to a couple of minutes. If you are in the middle of training, you will want to keep playing brief. For example, your dog has a tug rope they love to play with. You bring this with you every time you and your dog are training. When they try to perform a step for a trick, even when they don't succeed, you take the toy and play with them for thirty seconds. Then, you go back to training. At the end of your training session, play with them for a longer period of time as this will start to alert your dog that their training session is now over.

Take a Trip to a Dog Park

If you decide to take a trip to a dog park at the end of each training session, you will want to have another type of positive reinforcement during the training. This might be playing a game for a few minutes or a treat. You don't want to wait to give your dog the positive reinforcement until they are done with their training session as this will make training harder. Dogs learn through positive reinforcement. Therefore, the more you use it, the stronger your dog will react to training.

Give Them Attention

Give your dog positive attention when they complete a training step. You can pet them and tell them how great they are, give them a hug, or any other type of attention. During the session, you want to keep the attention brief. Once the session is over, spend more time with them, just as you do when you play with them during training.

Preparing Your Dog for Training

There is a lot of information to remember when you are getting your dog ready for training. It's important that you do everything you can to start on the right side of training as this will help your dog learn quicker. It will also give you and your dog the best possible training experience.

Below are a few tips to follow before you start training. While some of these tips are specific to obedience classes, most of the tips associate with home training and obedience school.

Prepare Any Documents

When you enroll your dog into obedience school, you will need to have the correct documentation. For example, you will need papers from your dog's veterinarian that proves they are healthy enough for obedience classes. This includes proof that your dog had their shots.

Bring Your Dog for a Check-up

Let your dog's veterinarian know you are going to start training your dog. They will do a check-up to make sure that your dog is fit and healthy for their training. They will make sure that they received all their shots and give you tips on how to help your dog through the training process.

Most veterinarian clinics have booklets about training that they will hand you. This is free reading material to help you through the process. They can also inform you of your dog's diet and if it needs to change to ensure they remain healthy throughout the training process.

Understand Your Dog's Attention Span

Plan a little play and rest time between your training sessions. How long a training session can go will depend on the age of your dog. For instance, a puppy shouldn't sit through a whole session for longer than five minutes without a break. An older dog can go for about ten to fifteen minutes without a break. However, if you are outside on a warmer day, you will want to plan more water and rest breaks, so you dog doesn't become dehydrated or too warm.

The breaks don't need to be long. You can schedule about a minute or two for each break. This means that you will train your puppy for two minutes, then give them a minute break. Train your puppy for another two minutes, and then give them a minute break. Finally, you will train your puppy for another minute before ending the training session. You can repeat this type of session a few times a day. Some people set aside a time to train their dog in the morning and then in the evening. Other people focus on training their dog every two hours.

Make Sure Your Dog's Stomach is Empty

Don't feed your dog too close to a training session as this can cause them to have an accident. Even though accidents can still happen, you will decrease the chances. Try to have your dog use the bathroom before the training session starts, especially if you are bringing them to an obedience class. Your dog is more likely to feel alert and pay more attention when their stomach isn't full.

Have Everything Ready for the Training Session

If you are going to use treats or toys to during the training session, have them ready to go before you start training. It might be helpful, especially if you are just starting training, to hide the treats and toys so your dog doesn't notice them immediately. It can be harder to get them to follow your direction when they are distracted by their favorite objects and treats. You are also free to bring any treats and toys with you when you go to any obedience classes.

Keys for Successful Dog Training

Throughout this book, you will learn about a variety of tricks to teach your dog. Some of these tricks will be basic, such as potty training, shaking hands, and sitting while other tricks will focus on advanced training. No matter what type of training your dog is learning, you always want to keep these training keys in mind.

- **Patience:** All dogs learn at different rates. Don't feel that your dog needs to learn as fast as your friend's dog. If they learn slower, it doesn't mean something is wrong. It simply means that they take a little more time to learn and this is fine.
- **Have realistic expectations:** Don't go into training with the thought that your dog is going to learn everything right away. They are going to learn the tricks one step at a time, and it can take a couple of days or longer for them to catch on. Lower your expectations so you won't feel disappointed as your dog will catch on to this feeling.

Plan ahead: If your dog doesn't understand what you are trying to teach them, break the steps down even more. I try to keep the steps simple for your dog in this book, but there are still ways to help your dog learn in smaller steps. The goal is to set your dog up to succeed. If you need to plan out the steps differently, do this before you start training.

Always show kindness: Always be positive and use positive reinforcement when your dog is learning. Dogs want to impress their owners, they are going to try their best, but also make mistakes. The more kindness you show them, the more comfortable they will become and willing to listen to your commands.

Generosity goes a long way: While you always want to watch how many treats your dog consumes in one setting for health reasons, you can show as much positive reinforcement as you want. There is nothing wrong with giving your dog extra love, toys, and praises.

Always remain positive: If you start to feel frustrated, stop training and take a break. Your dog will only have enthusiasm for training if they see you are enjoying the process.

Practice, practice, practice: You want to practice the tricks your dog is learning and knows often. Remember to keep all training sessions short, but you can hold several throughout the day.

Set training goals: How are you going to know what you want your dog to achieve without training goals? Before you start training, take time to establish goals. Keep them realistic as this will help your dog thrive.

Stay away from discipline: Harsh punishment can cause your dog to feel afraid and interfere with training. If you want to correct your dog's behavior, focus on positive reinforcement techniques.

Reward your dog effectively: Your dog will become more motivated when positive rewards are enforced. Take time to give your dog positive attention, play with them, and show them that you are proud of their accomplishments and what they are achieving.

Chapter 3: Tips and Tricks for Training Your Puppy

The most common age for people to train their dogs is the puppy stage. This is also known to be the hardest stage because their attention span is short, and they are easily distracted. To have the best possible training experience, you will want to work with these factors and not try to force your dog to change what is natural for them at their age.

Know Your Puppy Before Training Begins

Before you bring your puppy home, start to do a little research. Get to know the main parts of your dog's personality by reading about their breed. All dog breeds tend to have their own characteristics, but your dog will also have their own personality. For example, your dog's breed might be on the shy side, but you notice your dog tends to warm up to people quickly. You then learn that not every dog from this breed is the same. In fact, most dogs remain hesitant around people they do not know.

Once you get your new family member home, spend time with them. You want to play with them, talk to them, and observe their behavior. Dogs need time to play and be by themselves and this is the perfect time to watch your dog's behavior. You also want to observe your dog's behavior when you are playing with them and when they are eating.

While you want to know the house rules before you bring your puppy home, if you want them to become comfortable quickly you will not become critical of them immediately. You won't harshly discipline them. Instead, you will understand that your puppy is learning their new environment and focus on giving them affection. You won't allow them to run wild in your home, but you will need to be patient when enforcing the rules.

When your dog is comfortable, they are going to listen to you better. They will follow your rules and the relationship between you and your dog will continue to grow strong. If you focus on rules before affection, your relationship will suffer. They will become more fearful of you, which can cause them anxiety. This will make training harder for both you and your dog. To correct your dog, keep your voice soft and start using simple words, such as "no." Always remember when your dog listens you should praise them with positive reinforcement.

Within a couple of weeks, you will have a good handle on your puppy's behavior. Expect their behavior to change as they grow, because it will. They will grow naturally as they age, with training, and becoming comfortable with their living environment. Once you feel your dog is comfortable and you should start to understand some of your dog's body language, which will help you understand your dog throughout the training process. For example, you will realize when your dog needs a break and when something is wrong.

The best time to start training a puppy is when they are a few months old. Professional trainers often say the best time to start training your puppy is between twelve to sixteen weeks old as this is known as the "golden period" (Klinger, 2019). This is the time when your dog will understand praise means they did something well, making them more likely to repeat the behavior.

One of the reasons it is important to get to know your puppy before training begins is that it allows them to get to know you. They will also understand certain pieces of their environment, making training a little easier over time. Once you feel your dog is comfortable in their new home with you and are old enough, you can start training your puppy with the basic foundations, such as home base, potty training, sitting, and going to bed.

Book 2 - Tips And Tricks To Dog Training

Tips for Training Your Puppy

Puppies are quick to learn, but they can also cause a lot of stress in training. The key is to remember a few tips to make you and your dog's training experience the best.

Have Patience

Even though your dog is a quick learner, you will still need to practice patience. Puppies become distracted easily, they have trouble concentrating because of their short attention span, and they tend to be stubborn from time to time. This means that if your puppy doesn't want to focus on training, they are going to find something else to focus on.

You need to keep in mind that a lot of this is biological for your puppy. It takes time for them to learn how to read your behaviors and understand what you want from them. Don't become angry if your dog keeps becoming distracted during training. If this is frustrating, take a bit of a break and try again later. If you seem to have trouble gaining your dog's attention every time you are trying to start a training session, enroll them in an obedience school or talk to a professional trainer for advice.

Your Puppy Is at the In-Between Stage

Your puppy is not an adult or an infant. There are in-between stages when it comes to dogs. They are still developing mentally, physically, and emotionally and this can cause stress with training. At the same time, puppies can fool you. Because they are young, you may think they can't do a lot of tasks but think again. Puppies can do a lot more for themselves then they let on.

This age can also be tough for trainers because they struggle to know what their dog can handle and what it can't. The key to remember is that you need to take training one level at the time. Teach your dog the basics first, such as sitting, getting down, not jumping, laying down, and home base.

Train them on one trick at the time because they will become overwhelmed trying to learn more than one at once. Once they have a good hold on one trick, spend a couple weeks practicing this trick in your home and out in public. When they listen to you with very few mistakes, start training them on the next trick. Always remember to continue practicing the previous trick.

Your Puppy Has Fears

No matter how long you spend focusing on giving your puppy a comfortable home when they first come into your front door, they are going to be afraid. They may show you their fear and they may not. Some dogs try to hide their fear because they have a need to prove they are "fearless." Don't become concerned if your dog shows fear immediately. They are unsure of their new environment, you, and your other family members. They will need time to warm up.

During the puppy stage, they are going to act startled from time to time. This happens when they meet someone new or they hear a strange noise. You may see them jump, look around, sniff, and then continue on with their task. This is normal behavior for a puppy. When your dog doesn't seem to calm down and continues to show signs of fear is when you need to consider taking them out of the environment. Signs of fear include:

- Not eating or eating very little
- Not drinking water like they usually do
- Looking around the room
- Trembling
- Whining
- Trying to escape the area to avoid what is frightening them
- Diarrhea
- Unable to control their bowel movement
- Vomiting
- Panting excessively
- Salivating

If your puppy starts showing these signs and not letting up, it is time to take them out of the environment. Spend some time alone with your dog to comfort them and let them know that it is okay. Don't force them to go back into the environment if they don't have to. Chances are they will continue to feel afraid because something is causing them to feel this way.

If your dog is over five weeks old and continues to show these signs, they may have inherited fearful tendencies from one of their parents. In this case, you will need to talk to your dog's veterinarian or a professional trainer who can help you ease your dog's fears. This is important because your dog will remain fearful for the rest of their life if their fearful tendencies as a puppy are not taken care of.

Understand Your Puppy's Developmental Stages

In general, there are five developmental stages for puppies. Unless you have your dog from when they are born, you will not see all of the puppy's developmental stages. Once the puppy reaches their fifth stage, they are considered an adolescent dog.

Stage One: Neonatal Stage

Stage one occurs within the first two weeks after the puppy's birth. This is a crucial time for a puppy and their mother to bond. During this stage, it is essential that you check on the mother and her puppies to ensure they are healthy, and all puppies are eating, but don't focus on spending too much time with the puppies.

This is time for the mother and if the mother feels people are spending too much time with her puppies, she will become uneasy and want to move them somewhere else. The mother is the strongest influence for the puppy. They will begin to taste and touch right after they are born.

Stage Two: Transitional Stage

This stage starts when the puppy is two weeks old and goes until they are four weeks old. The puppy's behavior is heavily influenced by their mother. Therefore, you don't want to cause any stress on the mother. Even dogs feel overwhelmed by their new litter and are prone to aggression and depression easier. If your puppy feels this from their mother, they are more likely to show these characteristics later.

During this stage, the puppy will start to hear and smell. Their eyes will open, and they will start to stand, walk, bark, and wag their tail. At first, this won't happen a lot, but you will start to notice an increase in movement and sound coming from their location. Near the end of this stage, the puppy will see well, and they will start to show their teeth.

Stage Three: Socialization Stage

At this stage, puppies start to have longer stages of development. The socialization stage will start before the puppy is fully out of the second stage and it will last until they are twelve weeks old. During this stage, you want to make sure your puppy is getting all their social needs met, even if you don't bring them home until near the end of this stage.

There are several milestones during this stage:

1. Play time becomes important between three to five weeks old.
2. Puppies become more aware of their surroundings, people, and animals around them, and their relationships.
3. Puppies are highly influenced by their littermates between four and six weeks old. It is during this time they start to act more like a dog.
4. Starting at five weeks old, you need to make sure your puppy gets a lot of human interaction but start slowly. You don't want to overwhelm your puppy in the beginning because they will be a bit frightened by every new person they meet. Try to ensure all the interactions with humans are positive as this time sets the tone for how the dog is going to react to humans throughout their life.
5. Curiosity increases and puppies start to get into everything. It's always a good idea to "puppy proof" your home. They will also need more supervision during this time period so they don't get into anything that can harm them.

6. The puppy's senses are fully developed starting at seven weeks. However, they will continue to work on their coordination ability. This is always a fun time to watch your puppy as they start to stumble and see what their small body is capable of.
7. However, watch to make sure they don't hurt themselves as they will try to go up and down the steps and jump onto any piece of furniture. If they see other pets or their mom doing something, they will want to try too!
8. Around week eight is when you want to start house training your puppy. The first step should be potty training.
9. Week eight is when your puppy will start to show true fear and the signs previously discussed. Spend a lot of time comforting your puppy when they become afraid.
10. Week nine is a great time to start training your puppy how to act around people. You always want to start small. They should understand the word "no" by this point. If they don't, it's a perfect time to incorporate this word into your training.
11. If they don't respond to their name, start the training process so they know to come to you when you call. It will help ease you and your puppy into more training in the next phase.

Stage Four: Ranking Stage

This stage starts at about month three and goes to six months. This is when the puppy starts to understand ranking, such as dominance and submission. Because of this, you want to start training your puppy and establish the owner-dog relationship. The more they understand that you are the one who gives them orders, the easier training will go throughout their life.

During this stage, you want to make sure your puppy has time to play with other dogs. Allow them to play with dogs of any breed and size. As long as they won't hurt your puppy, they will start to thrive in understanding the ranking system.

One of the biggest challenges during this stage is chewing. In general, puppies chew on everything because they are teething. Just like teething bothers babies, it bothers puppies. Unfortunately, there is very little people can do to help ease any teething pain. At the same time, you don't want them chewing on everything. Because they take part in the action so often, it's best to train them not to chew on unwanted items at this stage.

One way to keep a puppy from chewing on something they shouldn't is to distract them. This works because puppies are easily distracted. While some people state this doesn't directly teach them not to chew, it is important to realize this is a temporary stage for a puppy.

You also need to be very careful when training your dog not to chew because there are items they can chew on, such as a dog treat, bone, toy, etc. Some people will give their dog a chewing toy when they catch them chewing on a pillow. Over time your puppy will understand they are to chew on their toys and not the pillow. Remember to give them positive reinforcement every time they start chewing on their toy over another item.

You can also take the object they are not supposed to chew on away and say "no." You don't need to sound harsh when you say no, just directly stating the word lets your puppy know that they shouldn't chew on that item. The key is you need to be consistent. Another tip for chewing is to play more with your puppy. When dogs are tired, they tend to stay out of mischief.

Stage Five: Into Adolescence

The adolescent stage starts at about six months and lasts until eighteen months of age. There are more challenges with this stage than the previous stages because dogs tend to challenge people more during this time.

Foundation Training

Foundation training is the basic training techniques. During the puppy stage, you will want to focus on potty training and home base. You will focus on more foundation training during their adolescent months, which we will cover in the next chapter.

Potty Training

Potty training your dog will take a lot of time and it will be stressful. It's also a process that people tend to find what works for them. There are several ways to train your puppy to go potty. One of the key factors you need to remember is you want to pick a spot the first time you work on potty training and always use this spot. This could be a pad in your house or a spot in your yard.

Your dog will roam around and use the bathroom in other areas outside once they are a little older. When they are a puppy, you want to focus on one spot. Do not use newspapers. You want to have something that is sanitary for your puppy and your home.

1. **Pick a spot:** You want to choose a spot to lead your dog to when they need to use the bathroom before you start potty training. You can mark this spot, so you remember where it is, or other members of your family know where to bring the puppy. If you allow your puppy to roam in the yard during potty training, they are going to become too distracted and forget to go...that is, until they come back into the house.
2. **Walk your dog to this spot every hour:** Bringing your dog to the same spot will explain to them where they should go. Only stay in the spot for one to three minutes. If your dog uses the bathroom, reward them with positive reinforcement immediately. If they don't, take them outside in another ten to fifteen minutes. Repeat this process until they use the bathroom.
3. Always remember what goes into your puppy will come out, and quickly. If they eat, take them outside when they are done. You will quickly learn that your puppy has a bit of a schedule when it comes to going potty. It will always help if you stick to a tight food and water schedule.
4. **Longer breaks start at twelve weeks old:** Around twelve weeks old, your puppy generally won't need to use the bathroom every

hour. You can increase this time slowly, starting with one hour and fifteen minutes.

Remember, there will be accidents—a lot of them. However, when your dog starts having fewer accidents, it means they are catching onto potty training.

Bed Training

1. Have your puppy face their bed. Point your finger toward their bed and say "bed." You only want to give the command once as you want to teach your puppy to follow the command when you only say it once. Consistency not only means you give them the same verbal command every time, but also the same number of times. For example, if you say the first time, "bed, bed" you want to say that every time. Your dog can become confused if they hear "bed" one time and then "bed, bed, bed" the next time.
2. Once your dog gets into their bed, even if they don't lay down, give them positive reinforcement.
3. If they leave their bed, repeat the command just as you did the first time. Again, if they get into the bed, reward them.
4. Follow these steps for two to three minutes or until you puppy is becoming too distracted. Remember, they are learning and have a short attention span. This is where your patience comes in.
5. Once your dog goes to their bed and lays down when they are standing in front of the bed with you pointing and giving the command, back your dog away further away from their bed. You don't have to leave the room, but you want to be at least a couple of feet away from the bed. Repeat the process.
6. Once your dog is comfortable going to bed from a few feet away, practice this trick from the next room. You will continue to back further away from the bed the more your dog listens. The goal is they will go to bed when you say the command once, no matter where they are in their home base. Again, always reward them with positive reinforcement.

Book 2 - Tips And Tricks To Dog Training

Home Base Training

Home base training is also known as perimeter training. This training teaching your dog where they can go and where they can't go. If you fear your dog running off because you don't have a fence or you don't want them going into certain rooms in your home, you will want to teach them this type of training. It is best they understand where they sleep (bed training) before you start their home base training as you can use their bed as the main home base point, meaning most of your training starts in this area.

Some people like to use the e-collar for this training because you can quickly alert your dog when they have gone too far without keeping them on a leash. However, you want to make sure your dog is old enough for the e-collar before this type of training. You can also use the clicker to alert them when they have gone too far. The key for the clicker is you want to be close to your dog, so you can give them a treat for coming back to their home base as soon as possible.

- Know where your dog can go and where they can't. Set up boundaries, such as cones or flags, to alert your dog when they have gone too far. It is important to stick to these boundaries. For example, can your dog go on the sidewalk? Can your dog go in the living room, bedrooms, and bathroom?
- Start by taking your dog around home base with the leash, clicker, or e-collar. When they pass the cones or flags, get their attention and give them the verbal cue of "no" or "home" and bring them back into home base. Repeat this step a few times until your dog stops going beyond home base as often.
- Bring their bed outdoors and place it in the middle of their home base area. Play with your dog, reminding them to come back to their bed when they pass home base. This step gives your dog a little more freedom and a specific location they need to go to when they pass the cones.
- Over time, you will notice your dog stopping near the cones. They might hesitate and then pass (remember, they are puppies and

curious), but they should quickly come back when you call them.
- Use the bed as their specific location until they start running to it naturally when they cross the home base line.
- While your dog is in their home base area, walk beyond their perimeter line. This is basically a test to see if your dog follows you, if they hesitate, or if they wait for you to come back. The goal is to get your dog to wait for you, but this won't happen right away. A puppy is most likely to follow you because they don't like being alone. You can save this test until they are in the adolescent phase, but you want to keep it in mind for when the right time comes.
- Another test for your dog, when they are older, is to watch them as they are freely moving around your yard or home. Of course, you want to be close, so if they go beyond their perimeter, you can call them back.

One key factor to remember is no matter how well your dog is trained, stay in your yard. When they are outside you should supervise your dog. This is especially important when they are still puppies. They may run to chase a car or go up to people when they are walking by. Furthermore, your dog is going to become excited and anxious when they see another dog approach.

Another tip is to wait for home base training until your dog is an adolescent. This can be a difficult training technique for puppies to learn. It should be taught near the end of the puppy stage or the beginning of the adolescent stage.

Tucker's Perimeter Training

Max got Tucker when he was eight weeks old. The first week, Max focused on getting to know Tucker and allowing him to explore his new home. Max didn't worry about Tucker running off as he was always with his dog outside, but one day when Max talked to the mailman, Tucker dashed out the door and into the street.

Max quickly got Tucker's attention by calling his name, but he didn't understand what Max was saying. So, he sat on the road until Max picked Tucker up and brought him back inside.

After talking to the local dog trainer who operated the obedience school, Max enrolled Tucker in the first class and received about home base training. Fortunately, Tucker knew to go to bed when Max told him to, so Max decided to purchase cones and place them around his yard.

He then put Tucker on the leash and walked him around the yard. Whenever Tucker tried to go beyond the cones, he took out the clicker and pushed the button. Tucker looked at Max, who would say "home" and bring Tucker back inside the cones.

For the first week, Max simply walked Tucker around for two to three minutes. He always used the clicker, told Tucker "home," and gave Tucker positive reinforcement when he came back to home base. Soon, Max noticed Tucker didn't walk beyond the cones as often. They would repeat this about four times a day.

During the second week, Max decided to slowly decrease the use of the clicker. Max's goal for training was to say "Tucker, home" and his dog would return home. Using the same training schedule as the previous week, Max used the clicker every other time. By Wednesday, he used it once a day. When Friday came, Max got rid of the clicker.

During the third week, Max decided to get rid of the collar. He wanted to play outside with Tucker without having to keep him leashed. While Max felt a bit nervous, he remained calm every time Tucker took a step outside of the perimeter line. Once Tucker heard, "Tucker, home," he ran back to Max, who responded with positive reinforcement.

In the fourth week of home base training, Max decided to stop walking Tucker around and play with him while they were outside. Max became pleased to notice that Tucker rarely left the perimeter lines. If he got close, he would hesitate. Max watched to see what Tucker did before saying, "Tucker, home." Even when cars and other people went by, Tucker stopped and watched them, but he never left his home base.

In the fifth week, Max decided to test Tucker. One day, Max ran outside of the perimeter. He stopped and observed Tucker's behavior. With the cones still up, Tucker stopped and looked at the cones. He then looked at Max and sat down. Max ran back into the perimeter, gave Tucker positive reinforcement and tried the test again throughout the week.

Because Tucker only left the perimeter lines a couple of times during his test, Max decided to hide when Tucker was outside. He was close to Tucker's hearing range, so could quickly call Tucker back if he went beyond the cones. Tucker never left his home base.

The next week Max decided to take the cones down and observe Tucker's behavior. Tucker ran around the yard and sniffed where the cones used to be but remained in the perimeter. While Tucker did go beyond the established perimeter lines now and then, he always came back when he heard, "Tucker, home."

Chapter 4: Tips and Tricks for Training Your Adolescent Dog

Before you start training your adolescent dog, you want to make sure you get to know them, especially if they are new to your family. If you have had them for a while and started training your dog already, you will continue training into this stage.

Adolescence starts when your dog ends the puppy stage at about six months old. However, some people (including trainers) still call a dog a puppy until they are about eighteen months old as they don't recognize adolescence in dogs. For the purpose of this book, the time your dog is considered an adolescent is between six and eighteen months old.

What You Need to Know About Your Adolescent Dog

Just as children go through an adolescent stage, most people believe dogs do as well. During this stage, your dog is going through changes, including their hormones. This can make them act in ways that isn't their normal behavior. It's important to remember to stay calm during challenges. You also need to remember that your dog's brain is not fully developed yet. Below are other important factors to keep in mind through the adolescent years.

Teething Is Almost Over!

While your adolescent dog is still teething, the worst part is over. They are starting to move out of their teething phase, and you don't have to worry about them chewing on everything. However, you will want to ensure they still have a variety of chew toys to keep them occupied when they feel the need to chew. Dogs love to chew on their toys and keeping toys handy will ensure they don't think of touching your furniture when you are not there. But your dog is growing and with that, their chew toys need to grow. For example, if you gave them a bully stick a couple of months ago and it took them one hour to work their way through it, don't expect them to take an hour now. Purchase bigger chew toys or watch them closely when they are chewing on their smaller toys.

Bond, Then Bond Some More

Bonding with your dog doesn't end in the puppy phase. While they trust you and are comfortable in their environment, they need constant reassurance that they are important to you. Think of your relationship with your dog as constantly growing and getting stronger throughout their life. Of course, if you recently got your dog you should still be focusing on establishing a tight bond.

Make sure you are spending time playing with your dog. Talk to them as dogs love to have a conversation with their companions. Don't save up all your time for when you are training them. Always make sure you set aside specific play time with your dog.

Keep Your Dog Social

Even though your puppy has now hit the teen years, you still need to focus on socialization. This is especially important if you just brought your new family member home and are unsure of how much socialization they have received.

Enroll your dog in obedience classes as this will help them become social with other dogs and people. Take your dog to the park a couple times a week, if not more, for exercise and socialization. If you don't have another dog at home, try to find your dog some friends they can hang out with.

Keep in mind, your dog is still going to act fearful when something or someone is new to them. Continue to support and comfort your dog when they are afraid, and they will become used to the situation or understand the person is okay to be around.

Book 2 - Tips And Tricks To Dog Training

Your Dog Is Not Going to Sleep As Much

Adolescent dogs don't need as much sleep as puppies. On top of this, their energy increases. Prepare yourself for busy days with your teen dog as they are going to keep you moving. Because of this, you want to keep them active throughout the day. Take them for a few more walks and play with them more. However, you don't want your dog to take part in any serious physical activity or sports yet. An adolescent dog's bones are still gaining their strength and they can easily break.

Did the Previous Training Fade Away?

If you taught your puppy to sit and you come to find that they don't sit as well as before, this is typical behavior for your adolescent dog. Mentally, dogs have a lot going on during this phase. They remember their previous training, but they can't seem to access everything they know right when they are supposed to. Continue to work on new training and practice the old training techniques you taught them so you can continue to build on them. Once your dog enters the adult phase, they will prove they remember everything you taught them.

Previous Dog Training?

If you just brought your dog home from the shelter, you won't know your dog's history. This means you won't know if they have had any type of training previously or not. Your best bet is to start with the basics or bring your dog to a dog behaviorist or trainer. They can give you information of what your dog may know, but without the cues your dog received from their previous trainer, it is hard to get your dog to follow your command.

Another option is you start training and find your dog catching on rather quickly over some types of training, such as laying down, but are slower to catch on to other types. This is a sign that they may have received training previously, but it wasn't consistent, or you aren't giving them the same cues.

Some people will tell you to start your dog's training from the beginning, just as you would a puppy. Even if they received previous training, you can retrain a dog in a different way, and they will follow your command.

Basic Training

If you just got your dog, you will want to start their training journey with sitting and home base training. If you received your furry friend as a puppy, you will want to continue training at this age. It's important to note that some of the training techniques I will discuss below can start at the puppy stage. Because there isn't a lot of time between the period you can start training the puppy at eight weeks and the period, they become an adolescent, the basic training mixes between the two phases. If your dog already knows to lay down, focus on the next type of training. If you are still working on teaching your dog how to sit, focus on this before moving on.

Come Training

Some people will teach them to come immediately after the dog learns their name. Other people will teach them after they learn to sit. The order you teach your dogs the basic commands depends on you. However, most trainers advise that learning their name and sitting should come first.

1. Start at your dog's home base by choosing a spot you want your dog to come to. Do your best not to move around until they show signs of understanding or you will confuse them.
2. Once you have your dog's attention, ask them to "come" and give them a hand gesture. Once they come to you, reward them.
3. Once they continuously come to you for a few training sessions, move around their bed, asking them to "come."
4. Next, place your dog in one room and go into the next room. While your dog can still see you, ask them to "come."
5. Once they do this repeatedly, move a little further away from their bed area and ask them to "come." At this point, you don't need to be in their line of vision but want to make sure you have their

attention.

6. Follow up "come" with an additional command your dog already knows, such as "sit." For example, you will ask your dog to come. Once they come to you, give them a little positive reinforcement. Next, you will ask them to "sit." Once they sit, follow up with more positive reinforcement.

Lay Down Training

1. Make sure your dog knows their home base before you start training. You will want to follow this step with every basic training method we discuss in this chapter. It is also helpful for some intermediate and advanced techniques, but not all.
2. Call your dog into their home base, usually where their bed is, by gaining their attention. If they understand "come" use that or call them by their name. When they come into their home base area, give them positive reinforcement.
3. You don't want to make your dog think you are telling them to go to bed. You want to keep "bed" and "lay down" separate because they are not always going to be in bed when you tell them to lay down. You might be at a friend's house, outside, in obedience school, or at the park. You want to get your dog used to laying down in any situation.
4. Ensure you have your dog's attention and say "lay down" with a hand gesture. You want this to be different than the gesture to "sit" or "come." It might help to get your dog to "sit" first and then get them to lay down. Like you did with sitting, you might need to gently push your dog down or lower a treat to get them to lay down fully. Remember to reward them, even for the smallest effort.
5. Once your dog is constantly laying down when you give the command, start practicing in other areas around your home and outside. You can even take them to stores that allow dogs and work on saying "lay down" when there are distractions around them.
6. Do your best to make sure your dog remains in the lay down position for a few seconds, As you continue to practice throughout the months you will increase the time, they remain in the lay down

position slowly. This will also be a part of most obedience classes.

At first, it is important that you don't force your dog to lie down for a long period of time. This can cause them to feel stressed as they are highly active dogs at this point. Even if they stay in the lay down position for a couple of seconds, make sure you give them positive reinforcement.

Case Example: Brick Learns to Lay Down

Tamara brought her dog, Brick, home when he was eleven weeks old. Over the last few months, she has taught him home base, come, and sit. She is now working on teaching Brick to lay down. As a positive reinforcement, Tamara uses treats. She also uses the clicker in order to get her dog to follow her command along with a verbal command.

Tamara starts the three-minute training process by calling Brick into his sleeping area. Once he comes, she gives him a treat. With the clicker in her hand, Tamara says, "Brick, sit." Brick responds by sitting, so she rewards him with another small treat. She then says, "Brick, lay down," and places her hand on his shoulder and brings a treat near the floor. As Brick is following the treat, she gently guides him down on the floor. Once he is laying down the whole way, she gives him a treat.

When Tamara stands up, her dog stands up and she tries the process again. This time, instead of placing her hand on the dog's shoulder, she uses the clicker and places the treat close to the floor. When Brick lays down the whole way, she gives him the treat. At this point, the three minutes are up. Tamara gives Brick some much needed attention as more positive reinforcement.

A couple of hours later, Tamara calls Brick into his sleeping area again. She repeats the training method with the clicker for another three minutes. Tamara and Brick go through this process four times a day. Within a week, Tamara is training Brick to lay down in other places around his home base. The following week, Tamara brings Brick to the park where they practice laying down when Brick gets excited about meeting a new person.

Book 2 - Tips And Tricks To Dog Training

Stay Training

Just like with lying down and sitting, your dog won't stay for a long period of time. They may only come and stay for a couple of seconds. Even so, you want to reward their efforts with positive reinforcement. You can work on slowly increasing their stay time over the next few months.

1. To start training, call your dog into their home brace area. This is a great opportunity to practice "come" training. Once your dog follows your command, reward them with positive reinforcement.
2. Tell your dog to "sit" but do not give them their positive reinforcement right away. Instead, place your hand on the top of your dog's head and say "stay."
3. Give your dog positive reinforcement by playing with them to get them back up again.
4. Repeat the sit command and then say "stay" and wait a couple of seconds to ensure that they stay. Reward your dog.
5. Continue this process and gradually increase the amount of time your dog stays.
6. Once your dog is sitting for about five to ten seconds, take a couple of steps back. Even if your dog moves, reward them for the effort.
7. Continue to take a couple of steps back from your dog once they stay for a short period of time. For example, if your dog stays for ten seconds after you take a couple of steps back, the next time that you focus on stay training, take a couple of more steps back. Eventually, you will leave the room and see if your dog stays. Don't keep your dog staying alone in the room for too long because they might start to worry. This is another trick that you will gradually increase the amount of time your dog stays.
8. Once your dog is comfortable with the "stay" task, you will want to add in other commands, such as lay down.

Case Example: Allie Learns to Stay

Amirah brought her dog, Allie, home a few months ago and started training her a couple weeks later. Now, Allie is in level two of a local obedience class, so Amirah is working on getting Allie to stay as this is what they are working on in class.

Amirah calls Allie into her sleeping area and gives her positive reinforcement by giving Allie a few seconds of attention. Amirah then stands back up and tells Allie to "sit." Once Allie listens to the command, Amirah tells Allie "stay" and holds up her hand in front of Allie's face as a hand gesture. After a couple of seconds, Amirah rewards Allie.

Once again, Amirah tells Allie to "sit" and "stay" as she raises her hand. At this point, Amirah decides to take a couple steps back. While Allie seemed like she was going to get up, she sat back down and stayed there until Amirah rewarded Allie with positive reinforcement.

After a few hours, Amirah called Allie back into her sleeping area and practice "sit" and "stay" again. This time, Amirah took several steps back, but Allie stood up and followed. Amirah rewarded Allie for her efforts with positive reinforcement and tried again, this time sticking to a couple of steps back. After a couple of times, Amirah moved four steps back, Allie didn't move until Amirah gave her positive reinforcement.

After a week, Amirah decided to add the command "come" after telling Allie to "stay." However, Amirah only said, "Allie, come," after taking about five to seven steps back. The next week, Amirah went into the next room as Allie stayed near her sleeping area. Allie didn't move until she heard her companion say, "Allie, come."

Get Down Training

It is easier to train your dog to get down when they jump up on something. Therefore, you don't need to start this training in a specific location as you want to catch them in the act.

1. Once your dog jumps up on something or someone, say "down" or "off" while you are showing them a treat in your hand.
2. When your dog gets down, reward them with the treat.
3. It is important that everyone gives your dog positive

reinforcement. For example, your dog jumps on your friend who walks in the door. You respond by telling your dog "off" and show him the bag of treats you have handy. Your dog gets down and you give him a treat. At the same time, your friend gives him some attention.

You will find it hard to always have treats handy whenever your dog jumps up on something they shouldn't. You can always use a different type of training, such as the e-collar, to train your dog to get down. However, you will still need to show some type of positive reinforcement, such as playing with your dog when they get down.

Case Example: Matt's Get Down Training

Donnie brought Matt home about four months ago. Since Matt knows how to sit, lay down, and come, Donnie has started to focus on telling Matt to "get down" when he jumps on people as they walk in the front door. Because Matt only jumps on people as they enter the door, Donnie decided to keep treats by the door.

Every time Matt jumps on someone, including Donnie, he grabs the treats bag right by the door and says "down." The problem Donnie runs into is that it doesn't work for him because Matt becomes interested in the treats and won't get down through verbal command.

Because Donnie can't keep Matt's "get down" training consistent, he needs to figure out a different way. After some research, Donnie decides that Matt is old enough of an e-collar.

Once the e-collar comes, Donnie places the e-collar on Matt and allows him to get used to it for the first week. When the day came for training, Donnie told Matt "down" when he jumped on him. When Matt didn't respond, Donnie pressed the button on the e-collar to send Matt the signal. Immediately, Matt got down and shook his head a little. Donnie rewarded Matt for getting down.

Within the first week, Donnie noticed that Matt started hesitating before he jumped up on someone. Donnie only sent the signals to Matt if he didn't get down after Donnie said "down."

Real Life Training

It's important to train your dog in a controlled and uncontrolled setting. A controlled setting is your home as you can control the area you train your dog, who is around you, and most distractions. The trouble with this is, if you are going to take your dog into certain stores, out to the park, or for walks, you can't just train them a controlled setting. You also want to train them in a real-life setting.

There are a lot of surprises that can pop up when you are in public with your dog. One of the greatest ways to prepare for these surprises is to practice training in public. To give you more of an example, let's look at how Roger trains his dog, Gidget at the local pet store:

After bringing Gidget home from the animal shelter two months ago, Roger is focusing on teaching his dog to stay. After training Gidget is at home for a couple of weeks, Roger decides to take her to the pet store so they can pick up more food, treats, and a new chew toy for Gidget.

While she is on her leash, Roger is looking at the dog treats when Gidget starts to move around. Roger looks down at her and says, "Gidget, sit." She looks up at Roger and sits down. Roger praises his dog, but notices that she is distracted by something else.

Looking in that same direction, Roger sees a larger dog a few feet away. It is at that moment Roger starts noticing that Gidget is acting a little strange. As he observes her, she seems to be getting anxious. The larger dog doesn't seem like a threat, but Roger knows that anything could have happened in Gidget's past.

Roger tells Gidget that it is okay, but she keeps her eyes on the larger dog, who takes a couple more steps toward Gidget. At that moment, she stands up and starts to head towards the larger dog with a slight growl. Gently,

Roger says, "Gidget, sit." When Gidget doesn't respond, Roger stands in front of Gidget, to try to block her view of the larger dog.

Again, Roger says, "Gidget, sit." Gidget sits, but then stands up before Roger can praise her. Roger repeats calmly, "Gidget, sit." Once she sits down, Roger quickly says, "stay," and holds up his hand. Gidget stays sitting and Roger praises his dog. By the time he is done giving her attention, the larger dog is gone.

As the story about Roger and Gidget shows, you never know what is going to happen in a real-life setting. This is one of the biggest reasons you want to do enough training at home that you can use different techniques. One of the techniques Roger used is called blocking and it is a popular trick to use when you don't want your dog seeing something or someone that can cause them to become anxious or aggressive.

What to Do With Aggressive Dogs?

Whether it is your dog that becomes aggressive or another dog, you want to try to stop the situation as quickly as possible. The best way to do this is by knowing the signs of an aggressive dog. Some of the signs include:

- A stiff body
- Growling
- Snarling
- Biting
- Ears are pinned back

If you notice someone else's dog acting this way while you are walking your dog, try to get your dog to go a different direction. You might need to distract your dog or block them from seeing the dog to get them to go in a different direction.

Beyond Basic Training

Near the end of your dog's adolescent stage, you may start teaching them tricks that aren't considered basic. Some of these are advanced tricks to help your dog act their best while other tricks are for fun. These tricks can be games, dog sports, hunting, or anything else that is interesting to you and you believe your dog will enjoy. There are over a hundred dog tricks available at your fingertips.

Unless you are training your dog for agility training or hunting, any training beyond the basic training should be strictly fun for your dog. If you start training them on a certain trick and they don't seem interested, move on to a different trick. With so many tricks, it is possible that your dog is not going to enjoy all of them. If your dog needs to learn them, then work on it like you did with basic training. If they don't need to learn them, it is important you don't force them as this can lead to a difficult training period.

Pulling on the Leash

Dogs tend to pull on the leash when they are on walks or not interested in staying in the area they are supposed to in the backyard. When dogs pull on their leash, they can hurt themselves or get loose and go missing.

The main reason they pull on the leash is because they have a place they want to go, but they are unable to get there because you don't have the same idea, or they can't go beyond a certain point. It doesn't matter what size your dog is, pulling on the leash can be dangerous. Therefore, it is always in your dog's best interest to train them not to do this.

One of the best tips for training a dog not to pull is to keep them interested in their surroundings. Take your dog for a brisk walk and change your direction frequently. This causes your dog to pay more attention to their surroundings as you are offering more for your dog to look at. When this happens, they aren't going to get bored or start focusing on a specific location and become determined to get there.

1. Once you get your dog walking briskly behind you, look to notice when they turn to follow you. When this happens, stop and give your dog positive reinforcement. A lot of people will use the clicker

during this training because your dog will associate the click sound to them turning to follow you. They will then associate the positive reinforcement to the click, bringing their training into a full circle. If you don't use a clicker, use a verbal command, such as "good," "yes," or "turn." Remember to use a verbal cue that you don't use for any other type of training.

2. After your dog comes to follow you without pulling on the leash, get a longer leash to see if your dog will become distracted and go the other direction. If your dog does, stop, get their attention, and then give them positive reinforcement for coming to you. This is a great time to incorporate the "come" command. If your dog follows, continue walking but give them positive reinforcement at the same time.

3. As time goes on, start choosing busier places to walk and walk a little faster. Repeat the training steps to ensure your dog understands that no matter what is going on in their environment or what they want to see, they need to follow you when you are walking.

A Digging Dog

Dog's dig for many reasons. First, some breeds feel the need to dig because it is instinctive. They were bred to dig for food and other animals. Second, dogs dig because they are bored, lack of exercise, wanting to cool off more during a hot day, or frustrated.

Before you start training your dog to stop digging, you need to understand why they are digging. German Shepherds, Beagles, Huskies, and Terriers are dogs where digging is part of their instincts. When it comes to this, it's hard to teach them not to dig. You might even find yourself questioning if you should try to stop them from digging because it is part of their nature. However, you also need to consider your yard, the yards of your neighbors, and other factors.

Typically, if your dog is bred to dig, they will still do so now and then as training is going to be harder for them. However, if your dog does this because they are bored or need a stress reliever, training will be easier, especially if you work with your dog and keep them busy if they are bored and calm if they are stressed.

If you have a dog whose instinct is to dig, you can think about getting a sandbox or some other type of digging area for your dog. During your training, you can focus on taking your dog into this area to dig instead of digging in various places within your yard. Of course, your dog will still need supervision to ensure that no inappropriate digging happens.

1. Cover a digging area for your dog with sand or loose soil. To make the area more appealing to your dog during a hot day, give the area protection from umbrellas or place the area in a shaded space.
2. Bury your dog's toys and allow them to start going on a treasure hunt.
3. Every time they dig in the digging area, give your dog positive reinforcement. Do not punish your dog for trying to dig in an inappropriate area.
4. If you struggle with getting your dog to stop digging in the wrong area, make it unattractive by adding rocks.
5. Supervise your dog whenever they are outside so you can lead them into the correct digging area and praise them.

If your dog is bored or needs more exercise, you can follow the training steps and increase your dog's activity level. For example, take them on longer walks, play with them more, give them interesting outside toys that they can play with. It is helpful to keep the outside toys separated from the inside toys, so they don't become bored with the toy quickly.

Tug and Release

Tug and release is a game that you should teach your dog before you focus on fetch. Use can use tug and release as a way to help your dog learn one of the steps in fetch.

For this game, it is best to have two identical soft toys or ropes that your dog can tug on when you are gently trying to pull it out of your dog's mouth. You never want to pull hard on a toy when it is in your dog's mouth because you can hurt them. It is possible for your dog to pull hard enough that their teeth become loose.

The best verbal cues to use are "grab it," "tug," and "drop it." You will also use the visual cue of waggling the toy to get your dog interested in the game.

1. Hold the toy in front of your dog but wait for them to look at the toy. Once they do, give them positive reinforcement.
2. Once your dog grabs the toy with its mouth, say "grab."
3. Let your dog pull the toy out of your hand while you are holding it. Make sure that you are holding it lightly, so your dog won't hurt themselves. Give your dog positive reinforcement.
4. Grab the toy and gently pull. Whenever you pull, exclaim with enthusiasm, "tug!"
5. If your dog doesn't drop the toy, bring out the identical toy and get him interested in his. Once your dog drops the toy in his mouth, say "drop."
6. Allow your dog to grab the second toy and repeat the process. Each time you get to the "drop" step, start to say it a little earlier as this will teach your dog to drop the toy when you say "drop."

There are dogs that will do everything they can to keep the toy in their mouth. They won't let go when you show them the identical toy. If this is the case, have treats on hand. When you say "drop" toss a treat on the floor and then praise them.

Teaching Your Dog to Play Fetch

Most people find teaching their dog to play fetch relatively easy. It is almost like an instinct to some dogs. It's easy to teach your dog to play fetch by focusing on positive reinforcement instead of the clicker, treats, or the e-collar. However, if you have a dog that likes to run off from home base, you might want the added security of the clicker or e-collar, especially if this is the way you are teaching your dog to stay at home base.

There are three main verbal cues you will use when teaching your dog to fetch:

1. "Get"
2. "Bring"
3. "Drop"

Some people like to use "give" instead of "drop." The problem with this is your dog can end up biting someone or scratching a little hand with their teeth. They won't mean to do this; they will be excited and have trouble containing themselves. Furthermore, if your dog likes to play a game of tug-o-war, they might think of this game when you are trying to get the ball, frisbee, or other item out of their mouth.

1. Find two toys that look exactly the same or purchase two of the same toys.
2. Keep one toy behind you where your dog can't see it and the other toy in front of your dog.
3. Tease your dog with the toy. You don't need to tease your dog for a long period of time, just for a few seconds or a minute. The point is to get them excited to run for the toy when you toss it.
4. Once your dog shows interest in the toy, toss the toy a few feet and say "get." You don't want to toss it very far at first because you want to get to your dog as soon as possible so you can praise them when they have the toy in their mouth.
5. This step is one of the hardest parts of teaching your dog fetch because they are not used to bringing the toy back. To do this, say their name and add "bring" to the end. For example, "Buddy, bring." When your dog takes a couple steps in your direction, praise them by grabbing the toy in their mouth and playing tug with them for a few seconds. You want to make sure you drop the toy before they drop it. You want the toy to remain in their mouth.
6. Once you get back to training, show your dog the identical toy from behind you. Start teasing them with the identical toy. Once

they drop the toy in their mouth, say "drop" and praise them before throwing the identical toy.
7. While they are running to get the identical toy, pick up the toy that was in their mouth.
8. Repeat the process.
9. Once your dog starts to understand the process of fetch, you will start to combine the words "get," "bring," and "drop" into fetch. This is when you throw the toy and you say "Buddy, fetch." When your dog has the toy in their mouth, tell them to "bring" it. When they come, tell them to "drop" it. At this point, you will praise them once they drop the toy before continuing to play the game.

Peekaboo

Not all dog tricks need to take a serious tone. Other than fetch, there are a lot of other fun games you can teach your dog, such as peekaboo. This is a trick where you have to think of safety and size of your dog because they will stand between your legs. Only teach your dog this trick if they are small enough to fit between your legs when you are standing. You also want to make sure your legs are far apart, but you are comfortable enough, so you don't lose your balance and end up stepping on your dog.

While this trick seems easy enough, and most dogs get it within a few days, going between legs can be frightening to some dogs.

1. With your legs wide apart, command your dog to "sit" and "stay" behind you.
2. With a treat between your fingertips, reach it through your legs and toward your dog.
3. Try to not let your dog receive the treat or positive reinforcement until they are halfway through your legs.
4. Once you see your dog and right before you give them the treat, say "peekaboo."
5. If they are scared, drop the treat between your legs and give them positive reinforcement for trying.
6. When you repeat the trick, you want to focus on saying

"peekaboo" earlier each time. The idea is that the cue for your dog to come between your legs is "peekaboo" and not the treat.
7. As your dog becomes more comfortable, you can move your legs closer together. Always ensure your dog can get through your legs without knocking you down or you stepping on your dog.
8. When your dog starts responding to the "peekaboo" cue, slowly decrease the times you use a treat to lure them between your legs.

Boing

When your dog is an adolescent, you are going to think about the various types of training for your dog. If you are interested in agility training, such as dog sports, you can start by teaching your dog to jump over a bar. Boing is also a great trick to teach your dog if you notice they have trouble with jumping or don't jump high enough. The best verbal cues to use with this trick are "boing" or "jump."

1. You want to lure your dog to jump. Most people do this by holding a treat between their fingertips, so their dog can see.
2. Get down to your dog's level by crouching and then spring up onto your tiptoes. When you jump up, raise your arm as this will get your dog to jump for their treat.
3. When you see your dog jump with you, say "boing."
4. Allow your dog to get the treat and give them positive reinforcement.
5. The next time your crouch down and jump up, raise your arm up a little higher. You want to gradually increase the length your dog jumps. If they are frightened of jumping, start lower, such as going from the floor to a little above their head.
6. You want your dog to learn to jump as high as possible, but not to the point they may hurt themselves coming down. If your dog is older, they may not jump as high. Senior dogs shouldn't jump as it can cause more damage to their joints, especially if they have arthritis.

Book 2 - Tips And Tricks To Dog Training

Time Out

While this trick is called time out, it isn't meant to be a punishment for your dog, though some people use this trick as such. One reason people teach their dog time out is because it can allow your dog to know that they need to calm down. Dogs are easily excited when there are other people around. Larger dogs can easily become overwhelming for young kids. If there is every moment where you want your dog to calm and show off their skills at the same time, this is a trick to teach your dog.

Time out is when your dog will lay on their side with their legs stretched out. To give them the visual cue, you hold your hand out flat, lay your hand in front of your dog, and flip your hand from palm up to palm down. The verbal cue is either "down lie" or "time out." Most professional dog trainers will refer to this trick as "down lie."

1. At first, this trick may be easier to teach your dog in steps they already know. For example, first tell them to "sit." After giving them some positive reinforcement, tell your dog to "lie down."
2. Once your dog is on the ground on all four legs, go to one side of your dog with a treat. You want to help your dog lay flat on their side. You can do this by holding a treat close to your dog's face and the luring it toward their shoulder. Their head will follow the treat.
3. You can start this trick with smaller steps. For example, once your dog sits on their hip, you give then a treat. Your next step will be to get them halfway laying on their side. Then, fully laying on their side with their legs stretched out.
4. Once your dog is laying flat on their side, say "time out" and give them a treat.
5. When your dog starts to react to the treat, add in the visual cue. Your goal is to get rid of using the treat and use the visual cue instead.

Roll Over

Like time out, roll over is another trick that you can teach your dog in smaller steps. You can also start with using the tricks your dog already knows, such as sit, lie down, or time out. The visual cue you will use is pointing toward your dog and making a circle with your finger. The verbal cue you use is "roll over."

1. Command your dog to sit.
2. Kneel in front of your dog with a treat between your fingertips.
3. Command your dog to lay down.
4. Command your dog to time out.
5. Lure your dog to roll over half-way by placing the treat by their nose, moving it towards their shoulder, and then slowly to the other direction. For example, if your dog is lying down facing east, you want to move your treat west. Your dog will follow the treat, making them go into a belly-up position and then reward them.
6. Start your dog back at the beginning and then practice having them roll over slowly, going a little farther to complete the roll over trick each time. Give your dog the treat and then positive reinforcement each time.
7. Once your dog fully rolls over with treats, introduce the visual cue. Keep using the treats at first. When your dog starts to roll over by following the visual cue, slowly decrease the treats you use to get your dog to roll over.

Hunting

Near the end of the adolescent stage, or at the beginning of the adult stage, most people start to train their dog for hunting. There are several breeds of dogs that are great hunters. In fact, it is often a part of their instincts and they are happy when they are hunting.

If you decide to get a dog and teach them to hunt, you want to look into getting an e-collar. There are specific hunting e-collars that will reach your dog to a mile in length. They can be a great tool as you can train your dog

Book 2 - Tips And Tricks To Dog Training

to go back home or come back to you when you send them the signal.

1. The first step you want to do is map out your hunting route. You want to make sure that you have a clear route from your home or vehicle to where you will hunt. You will also want to plan out the route you will take in the woods while you are hunting. If you don't know the woods well, mark your route in various places, so you can be sure to learn it as well.
2. Teach your dog the hunting route. Take them along the path a few times and let them see where your hunting base is. Allow them to get used to the smells of the route. Take your time so they become comfortable along the route.
3. Train your dog to return home. This will ensure that no matter what happens, especially if they go beyond the e-collar range, that they won't get lost. However, it is always best to have your dog in your sight or give them a signal as soon as you lose sight of your dog. Dogs can run fast, and it won't take them very long to get beyond a mile.
4. Retrain your dog on hunting every year before you take your hunting trip. Repetition is key, and while your dog will remember everything about hunting, it will help when it comes to the route. Things can change along the route in a year and if a tree falls or something else happens, your dog can become confused along the route. Furthermore, the smells are not going to be the same from year to year.

Chapter 5: Tips and Tricks for Training Your Adult Dog

Around eighteen months is when your dog enters adulthood. They will remain in this stage until they become a senior dog, which is anywhere between six to ten years old. By this age, you should be done introducing basic training, but still working on building the skills.

For example, you will teach your dog to stay longer over a period of time. Of course, if you just brought home your dog from the shelter, you will need to start with the basic training discussed previously.

While you will focus on the training tips discussed previously, your adult dog can train for a longer period of time. Instead of two to three minutes, you can go about four to six minutes. At the same time, you may need more patience. An adult dog has spent over a year or more going about their life a certain way, whether they were trained or not.

For instance, you got your dog from the animal shelter. One of the first traits you notice about your dog is they pee on paper they find lying around the house. You assume that this is because the previous owners allowed their dog to pee on newspaper. Therefore, you need to house train your dog to use the bathroom outside instead of paper they find around the house. This training can take longer than it would for a puppy because they have gone to the bathroom by using paper for a period of time.

Keeping Your Adult Dog Healthy

Focusing on keeping your adult dog healthy is a little different than a puppy or adolescent dog. First, you need to watch their weight more. Another factor is you have to make sure that you follow your dog's veterinarian's orders if they tell you to start your dog on any type of medication or vitamins. The more you focus on keeping your dog healthy and fit, the longer your dog will live. Here are a few tips for keeping your adult dog healthy:

Make Sure You Are Feeding Your Dog the Right Food

Diets for adult dogs are different from younger dogs. You may need to look into special diet food to keep your dog's weight under control. Because adult dogs aren't as active as puppies or adolescents, they tend to gain weight a little faster. This causes a lot of health problems over a period of time and will shorten their lifespan.

If you aren't sure what food is best for your dog, talk to your dog's veterinarian. This is especially important if your dog starts to develop signs of a disease or illness. While this specialty food is a little more expensive than common name brands, keeping your dog healthy and extending their lifespan is worth it.

Include Healthy Activities Into Your Dog's Schedule

Making sure your dog stays active is something you want to do throughout their life. However, as your dog ages, you will need to spend more time making sure your dog remains active. During this age, you can start to teach them how to play fetch, if you don't during their adolescent years. You can also teach them various games to play to keep their mind and body active.

Track Your Dog's Weight

Most veterinarians will start to tell you to track your dog's weight once they reach adulthood. This is because your dog can lose or gain weight rapidly. There are a lot of reasons for this and if you find your dog losing or gaining weight at a fast pace, you need to bring them in for a checkup. However, if they start losing weight at a slow pace, you shouldn't have to worry unless they begin to lose too much weight.

Increase Your Visits to the Veterinarian's Office

When dogs are young, most veterinarians only want to see them annually. Once your dog reaches adulthood, they will start to request more frequent visits. This might start at twice a year and then go to three times a year when they get closer to reaching the senior age. Make sure you follow the veterinarian's advice with how often they want to see your dog. If any type of health problem arises, you can catch it quickly and take care of it to increase your dog's lifespan.

Adult Dog Training Techniques

Before you start training your older dog, your dog needs to be comfortable in their environment. They need to know that you are there to help them and give them what they need to have a healthy and happy life.

If you recently received your dog from an animal shelter, you don't know anything about your dog's past. However, it is possible that your dog comes from a negative past where they were abused and neglected. This is going to affect your dog mentally, emotionally, and physically. It's important that you take everything slowly when bringing an adult dog home because they can have years of heartache and pain built up inside of them.

You never want to react harshly in any way, no matter what your dog does. You need to give them time to adjust before you start training. Even if they aren't well potty trained, be patient and kind with their accidents. The more you show your adult dog love and compassion, the stronger the bond will form. This won't help them get over any trauma from their past, but it will help them know that they are in a much better home.

Your dog's past can be the biggest challenge when it comes to training. If they dealt with abuse, they will have a hard time following commands. They might be more interested in watching your every move to make sure they can protect themselves because they are still learning how you are going to react and treat them.

If you start to feel frustrated, your dog is going to pick up on this and become increasingly anxious. You should only train a dog, especially older dogs who could have lived through trauma, when you are happy and calm. The moment you start to feel this slip away is the moment you need to cease training. Ensure you always end it by giving your dog positive reinforcement.

If you feel your dog is really struggling and shows signs of anxiety, depression, or fear, consider taking them to a dog behaviorist. They will help you and your dog overcome any of their trauma so your dog can start to heal and lead a healthy and happy life.

Training Tips for Older Dogs

Age Doesn't Matter

No matter how old your dog is, they can learn just as much as a puppy. Many people believe that an older dog cannot learn new tricks; this is a myth. Dogs love to learn and please their owner at any age. While you will need more patience and teach your dog at a slower pace than a puppy, your adult dog will quickly thrive in their new learning environment. You will be proud to show off all of the new tricks your dog knows.

Use a Crate

It's not always a favorite for people because they feel a crate cages their dog like a jail, but, to your dog, it can be a comfortable place to sleep and relax. This is especially true for older dogs who like to make sure they have a sense of security around them.

Build Your Dog's Confidence

If you recently brought your older dog home from a shelter, you want to focus on building their confidence before you start training them regularly. Dogs are going to focus more on their training and believe they are pleasing their owners if they have a higher self-confidence. Most people don't think about a dog's mental state, but how they feel about themselves and their surroundings is one of the strongest indicators for their success. Your dog is always going to try to please you, but by feeling more confident in themselves and their abilities, they will believe you are proud of them.

Back Up

Back up is when you teach your dog to walk backwards. This is a great trick if you find your dog in the way or you want to practice teaching them a new trick to keep their mind and body active. Most dogs understand this trick fairly quickly. The verbal cue you will use is "back" and the visual cue is waving your hand in the "leave" or "go away" motion.

If you want your dog to back up in a straight line, you want to set up a space where your dog will back up. This could be placing cones down the center of a hallway and allow your dog to walk between the cones and one side of the hallway.

You can even add to this trick. If you want your dog to learn agility training have your dog back up while weaving between the cones. However, this will come later as you want your dog to start by backing up in a straight line and then backing up between the cones. Remember, you always want to take little steps in training.

1. Stand at the entrance of the hallway. Have you back toward the hallway and your dog in front of you.
2. With a treat, lure your dog to follow you a few steps into the hallway. You will want the treat close to their nose.
3. After a few steps, stop, and take a few steps forward to make your dog back up. You can help your dog do this by putting the treat above their head and slowly moving it behind their head. If your dog turns around, give them a treat for effort. If your dog follows the treat by taking a couple steps back, give them the treat by tossing it between their front paws. This will make them back up a bit more.
4. Once your dog backs up with the treat, add the verbal cue of

"back" when they start walking backwards.
5. Slowly decrease the treats and focus on using the verbal cue. However, have a treat in your hand for positive reinforcement.
6. Add in the visual cue of waving your hand when you say "back."
7. If getting your dog to back up in a straight line is important, you want to practice giving them more space once they consistently back up five to seven steps. Remember to widen their path little by little. Don't take down the cones immediately. Instead, move them over an inch or two each time.

Case Example: Frankie Learns to Back Up

Frankie is a poodle near the end of her adult years. Her owners, Barb and Tony, have become concerned over Frankie when she is in the kitchen. While she has always been in the kitchen when the couple is cooking, Frankie's veterinarian told the couple that she has early stages of arthritis, causing her to walk slower. It's come to the point where the couple almost trip over Frankie in the kitchen because she can't move out of their way quick enough.

After talking to Frankie's previous trainer from her obedience class, they tell Barb to teach Frankie to back up, so she isn't in their way when they are cooking. Barb sets up cones in the middle of the kitchen and focuses on one side to teach Frankie to "back up." Starting at the entrance of the kitchen, Barb takes a few steps into the kitchen. She knows Frankie well enough to know that her dog will follow her.

Once Frankie is near the stove, Barb holds up the treat between her fingertips next to Frankie and starts slowly moving it. Frankie takes a step back, so Barb tosses the treat between her dog's front paws. Frankie takes another couple of steps back, eats the treat, and receives positive reinforcement from Barb.

When Barb tries to get Frankie to back up a little more with another treat, Frankie lays down on the floor and looks up at Barb. Knowing that Frankie is tired, or her joints are starting to bother her, Barb rewards Frankie for her effort and stops training until later.

When Barb tries to train Frankie again, her dog comes into the kitchen, but then immediately lays down. Barb gives Frankie the treat and then comforts her dog as she knows there is something wrong.

After talking to Frankie's veterinarian, Barb learns the best step to take is to train Frankie on carpet or rugs as hard floors are bad for Frankie's joints. Therefore, Barb starts to train Frankie in the living room instead of the kitchen. To ease more on Frankie's joints, Barb only does one round of training twice a day. Within a couple of weeks, Barb has taught Frankie to move backward and out of the kitchen with the verbal cue "back."

Frankie's story gives you an idea of what obstacles training can bring. While Frankie wanted to please her owners, she was also trying to tell Barb that something is wrong by laying on the floor of the kitchen. Hard floors do cause pain for dogs who are showing signs or have arthritis. It is important that you listen to your dog and watch how they are acting during training.

If you feel that something is wrong with your dog, stop training and contact your veterinarian. They will take time to discuss any type of problems and give your dog a check-up to see what they are trying to tell you. While most people think that their dog is going to remain healthy until later in life, many diseases can show themselves in your dog during the adult stage.

Stand on a Platform

If you're training your dog to become a show dog or for agility training, it is important you focus on training your dog to stand on a platform. Of course, you can also train them in this trick for the fun of it. Once your dog stands on the platform, you want to make sure they wait for you to give their next cue. Therefore, if you are doing this for fun, always remember you need to give your dog an ending cue.

Your dog's platform needs to be an elevated flat surface they can jump or

Book 2 - Tips And Tricks To Dog Training

climb onto without much trouble. If your dog is having trouble going up and down stairs, you will want to use a ramp to help your dog reach the platform. You can also think of using a specific rug as a platform for senior dogs.

The visual cue for this trick is pointing to the platform. The verbal cue is often "place," "table," or "hup."

1. Get your dog to get to know their platform. Place a few treats around the platform and allow them to sniff the platform, eat the treat, or stand on the platform. The second your dog touches the platform, reward them with verbal praise.
2. Don't start training them to step up on the platform or place your dog on top of the platform. They need to get used to the platform in their way. This can take a few sessions before they are ready for training.
3. To lure your dog onto the platform, use treat and verbal cues. Hold a treat between your fingertips and start by your dog's face. Slowly move the treat onto the platform. Once they touch the platform, say "place" and give them their treats and positive reinforcement. Mark the place they touched, so you can make sure they go a bit further onto the platform with each session.
4. Once your dog is halfway onto the platform and comfortable with the verbal cue, add in the visual cue. Point to the platform when you say the verbal cue.
5. Once your dog starts to jump onto the platform, toss the treat away from the platform. Repeat this process for a few seconds.
6. When your dog starts following your verbal and visual cues, slowly decrease luring them with a treat.
7. When your dog stays on the platform for a few seconds, tell them to "sit." Once they sit for a few seconds, toss the treat.
8. Move the platform around your home so they become familiar with the trick and not the location. This is necessary for show dogs as they will be in different areas when needing to stand on a platform.
9. When your dog starts sitting on the platform, take a couple of steps

back. The point is to keep your dog on the platform until you give your finishing verbal cue, which can be "down" or "come."

Touching a Target

The target your dog will touch with their nose is any type of flat and safe target. You can use a target, round lids, sticky notes, paper, or squares from the carpet. You want the target to be medium sized and something that your dog won't associate with a toy. You can also train your dog to touch the target with their paw.

The visual cue you will use is pointing to the target. If you decide to train your dog to touch the target with their nose and paw, you want to create two types of visual cues. For instance, you could point to the target with your finger for nose touching and point to the target with your hand held out and your palm aimed at your dog for paw touching.

The verbal cue you will use for targeting is "target." Again, if you are training your dog for nose and paw touching, you will want two verbal cues. For example, you can use "target" for nose touching and "touch" for paw touching.

1. Decide if you will teach paw or nose touching first, or only one.
2. Choose a target that will interest your dog, but not to the point they will want to play with the target. You want them to focus on the target.
3. Test your dog's interest with the target by tossing the target on the floor. Make sure you have your dog's attention.
4. If your dog looks at the target, give them positive reinforcement.
5. Leave the target on the floor and encourage your dog to walk close to the target. After they take a few steps or walk to the target, give them positive reinforcement.
6. When your dog touches the target for the first time, give them more positive reinforcement than the last few steps. You might do this by playing with them for a few seconds with their favorite toy or giving them a treat. Focus on this step for a few sessions.

7. Now, limit your dog's positive reinforcement for touching the target with their nose or paw. Remember to just focus on one. For example, your dog touches the target with their nose, so you say the verbal cue "target" and reward them. Focus on this step for a few sessions.
8. With the target in a certain spot and your dog a few inches away, say "target," and point to the target. When your dog touches the target with their nose, give them positive reinforcement. If your dog doesn't move to the target easily, reward your dog each time they move a bit closer to the target.
9. Once your dog is touching the target with their nose after your cues, start moving the target around the house and continue to practice.

Case Example: Minnie and Her Target

Minnie loves to learn new tricks. As a way to keep her older dog's mind active, Minnie's companion, Trisha, found the "target" trick. After reading the steps, she decides that her dog can easily learn how to touch a target with her nose. For the target, Trisha uses a yogurt lid.

In her dog's sleeping area, Trisha tosses the lid onto the floor. Minnie, who was playing with a toy, looks at the lid and quickly becomes curious about what just fell in her sleeping area. Slowly, Minnie gets up to sniff the new lid. Trisha rewards her dog with positive reinforcement and a treat.

For the next session, Trisha moves the target to a specific location in Minnie's sleeping area. With Minnie across the room, Trisha says "Minnie" to get her attention. When Minnie looks, Trisha waves to motion Minnie to "come." Minnie follows the command and Trisha rewards her dog with positive reinforcement.

Trisha then sees Minnie look at the lid, so Trisha says "target" and rewards Minnie with a treat. When Minnie takes a step toward the lid, Trisha responds with "target" and gives Minnie another treat. Each time Minnie takes a couple of steps closer to the target, Trisha gives Minnie positive reinforcement.

Once Trisha says "target" and Minnie responds by going to the target, Trisha starts using the visual cue of pointing. After following his for a couple of sessions, Trisha focuses on getting Minnie to touch the target with her nose. Trisha says "Minnie, target" and points to the target. Whenever Minnie touches the target with her nose, Trisha rewards Minnie with positive reinforcement. Over a few sessions, Minnie always responds by touching the target with her nose.

Cleaning Up Toys

No matter what age a dog is, they can leave a mess with their toys. As long as your dog understands fetch and that their toys go into a box, they can clean up their toys with a little encouragement.

If your dog's toys aren't already in a specific box, place them in a box that can store all of your dog's toys. Let this box full of toys that sit in your dog's sleeping area for a while and take out toys to play with. This will get them used to the fact that their toys are in this box. Remember, clean up your dog's toys at least once a day so they have to retrieve their toys from the box over and over.

The visual cue will be pointing to the toy box. The verbal cue will be "clean" or "clean up." You can also use "pick up."

1. When you start training your dog to pick up their toys, take all the toys out of the box and place one in the box. Ask your dog to fetch the toy. If they are confused, walk them to the box and ask them to fetch. Once your dog picks up the toy, set your hand on the box and ask your dog to give the toy to you.
You may need to use the phrase "drop" like you do during fetch. When they drop the toy, place it into the box and reward your dog. When you reward your dog, give them a treat by holding it over the box.
2. Teach your dog to drop the toy in the box. Once your dog has the

toy in their mouth, point to the box. If they drop the toy outside of the box, give your dog attention for trying. If your dog drops the toy inside the box, give them a treat and attention.
3. Take the toy out of the box and ask them to fetch it to practice dropping the toy in the box.
4. Once your dog is dropping one toy in the box continuously, add another toy.
5. Include the verbal cue with your visual cue. Point the box and say, "fetch, clean up." When your dog picks up the toys and places them in the box, reward them with a treat and attention. If they don't drop them inside the box, give them attention for their effort.
6. Once your dog is cleaning up two toys, add in more toys.
7. When your dog is consistently placing the toys in the box, drop the "fetch" and just say "clean up."

Catch a Ball

To catch a ball, it is best to teach your dog to fetch and play tug and release first. These are often known as building blocks to a game of catch. The point of this game is for you to toss a toy, such as a ball, into the air and your dog catches it in their mouth.

It's important to note that not all dogs like to play catch. If you find that your dog tends to move away when you toss an object instead of trying to catch it, they might be too afraid to catch an object in their mouth. Don't force your dog to learn catch.

For this trick, you want two identical balls. Practice tossing the balls gently before introducing the balls to your dog. The verbal cue for this trick is "catch."

Introduce one ball to your dog by rolling it on the floor by them. Keep the other ball in your hand. Your dog is going to become interested in the ball rolling their way. As the ball is rolling, you can build your dog's drive to get the ball by saying "get" just as you would during fetch. Once your dog is playing with the ball, give them positive reinforcement.

Take the ball your dog is playing with and hold it up to them. Toss it to them gently and say "catch." You want to toss it gently toward their mouth. The more the ball ends at your dog's mouth, the more confident they will feel about catching the ball. When your dog catches the ball in their mouth, give them positive reinforcement and a treat. If they let the ball drop to the floor without catching, give them positive reinforcement for effort.

Once your dog catches the ball a few times in their mouth, ask them to "give" the ball back to you. This is when the second ball comes in handy. The second your dog drops the ball from their mouth, give them positive reinforcement and gently toss the other ball in their direction.

Once your dog is trying to catch the ball confidently, start backing up by taking a step each time you toss the ball.

Chapter 6: Tips and Tricks for Training Your Senior Dog

Depending on your dog's breed, the senior age will start between six to ten years of age. If you just brought home your dog, you will want them to get used to the environment before you look at training. Instead of letting your dog take a week to get comfortable, give them a couple of weeks. You always want to allow a senior dog more time because they are starting to slow down in general.

Furthermore, you never know the background of your dog. If they are older and come from a shelter, they could have a lot of traumatic experiences in the back of their mind. This means they will need more time to get used to their new environment and become comfortable around their new family.

People often feel that you need to have more patience bringing home a senior dog than you do with a puppy. The main reason for this is because puppies adjust quickly since everything is new and exciting for them. When it comes to a senior dog, changes can cause a lot of anxiety and be overwhelming.

This can also cause them to have frequent accidents, even if they are potty trained. Don't become frustrated or angry with your dog if they have accidents. Understand that it is part of their system when things are changing, and they are feeling overwhelmed. Comfort your dog and let them know that you love them.

Keeping a Senior Dog Healthy

You are going to spend more time worrying about the health of your old dog than you are at any other age. Many people feel that a senior dog should see the veterinarian at least four times a year. If they start to have health problems, this can increase. Here are some other tips to follow when it comes to keeping your older dog healthy:

Pay Close Attention to Their Diet

You always want to pay attention to your dog's diet. You want to purchase nutritious food for their age that gives them all the nutrients they need. However, this becomes more important as your dog age and lifestyle. Senior dogs will benefit from a diet that has fewer calories and less fat.

You want to focus on vitamins and other minerals. This changes because senior dogs are less active than younger dogs. It's also important to make sure there is fiber in your dog's diet because senior dogs are more prone to constipation. Many people like to feed their senior dog some fruits and vegetables instead of just hard or soft dog food.

Dental Hygiene

Senior dogs can easily start to lose their teeth. It is important that you take your dog in for teeth cleanings, which is something that your veterinarian can do. Keep up with a check-up every six months and inspect your dog's teeth and gums at least once a week.

If you notice something doesn't look right or your dog has a loose tooth, call your veterinarian immediately. This can cause your dog pain. Furthermore, your dog can get food stuck in the hole in their gums and this can lead to an infection.

Make Sure Your Dog Gets Exercise

Your older dog is not going to pay as much attention to exercise. You might find your dog, especially if they are not doing well, not wanting to exercise. Do your best to get your dog up and walking. A larger dog's exercise should go beyond a walk around the block or take them around the block a couple of times a day. However, your smaller dog might feel like a walk around the block is a bit too much.

If you just brought home your dog and you are not sure if they are used to exercise, you want to start slow and gradually increase the amount of exercise they get in one day. Remember to notice how they handle hotter days. If your dog can't be outside for long on a hot day, make sure they get more exercise indoors. Walk them around the house or play with them more.

Provide Your Dog with Special Accommodations

Your older dog will have health issues. One of the most common problems for older dogs is arthritis. This is a painful disease and you want to do more than give your dog medication to help ease their discomfort. For example, dogs with arthritis benefit from towels, blankets, and other soft bedding. If you have stairs they need to climb, think of installing a ramp for them. Any hard floors that they need to walk on should have some type of rugs or carpeting for them.

Training Techniques for Senior Dogs

The most important factor to keep in mind with your senior dog is how much they can handle. There are tricks that you won't want to teach your older dog, such as agility training, hunting, or any tricks that can cause them to become tired quickly or hurt themselves. For example, if your dog doesn't climb stairs well, you will want to keep this in mind when training.

Senior dogs can learn the basic tricks, such as home base, sitting, and lying down without too much trouble. While you can usually train them for about ten minutes at a time, if you feel your dog is becoming tired or it's a hot day, you need to shorten the time and allow them to relax.

If you would like professional help with training your senior dog, think about enrolling them in obedience classes. Obedience classes do not have age limits. Any breed of any age can start learning tricks through these classes.

Training Tips for Senior Dogs

The training tips for older dogs discussed in the previous chapter need to be kept in mind when you are training your senior dog. At the same time, these dogs are near the end of their life and require extra attention when training. Here are some other tips to keep in mind.

Watch the Repetition

When you are teaching a younger dog tricks, you are told that repetition is key. This is something you don't want to do with senior dogs because of their joints. Repetition will cause any arthritis to flare up. It can also cause flare ups even if they don't show any signs of arthritis yet. This means you won't want to make your senior dog sit seven times in a row. Instead, have them sit a couple of times, or once if it makes you feel more comfortable, and then give them a bit of a break.

If you are training your dog to sit or lay down and they don't want to respond, don't get frustrated. The trick might cause them discomfort of pain. Instead, stop worrying about training, make your dog comfortable, and get them into the veterinarian for a checkup.

Shorten the Training Time

It's fine to train older dogs for then minutes at the time, but your senior dog may not handle this amount of time too well. If you start to notice your dog becoming tired or uninterested in training, start shortening the training time. Observe your dog during training and pay attention to when they start to get more sluggish and tired. If it is about four minutes into training, focus on training for only a couple of minutes.

Use Verbal Cues and Hand Gestures to Train Your Senior Dog

It's fine to use an e-collar or clicker for training a younger dog, but to train your older dog, you should stick to training signals, or verbal cues or hand gestures. If your dog is going blind, use verbal cues, and if they are going deaf, use hand gestures. Even if you used an e-collar when your dog was younger, the shock can cause them to move quickly or in a way that can cause more pain in the joints.

Train Your Dog on Soft Surfaces

Because of your dog's arthritis, it is best to train your dog on carpet, rugs, or other soft surfaces. Furthermore, training on a soft surface is more comfortable for your dog's paws and bones.

Which Hand?

The tricks in this chapter can be taught to your dog sooner. These tricks are ones that are easy to grasp and don't require a lot of time from your older dog. However, you should only focus on teaching your senior dog new tricks if they are capable of accomplishing them. If you start training your dog and find that they become tired or stop focusing on the step and lay down, your dog is telling you that it's not time to learn this trick. While you want to keep your dog's mind going throughout their life, it is important you don't push your dog to take on a trick that they don't feel like doing.

Which hand is a simple trick that most dogs enjoy because they focus on picking the hand with the treat in it. The challenge is you want your dog to place their paw up on your hand, telling you which hand they pick.

This trick is often used for police and military dogs because it focuses on a dog's scent. It will help them build up their scent skills by focusing on the area the scent is coming from.

At the beginning, your dog might need a little hint to let them know which hand the treat is in. At the same time, if your dog is losing their smell, you will want to give them a little hint to show them which hand the treat is in. You can do this by gently shaking your hand a bit.

1. Get down to your dog's eye level and place a treat in your hand. Show your dog the treat.
2. Close your hand to make a fist.

3. Ask your dog "which hand?" and command your dog to sit. Your dog should sit facing the hand with the treat.
4. Open your hand with the treat and allow your dog to eat the treat. Don't use the same hand each time. You want to switch it up, but make sure that they see which hand the treat is in to start.
5. After a few successful sessions, stop cueing your dog to "sit" and just ask "which hand?" Your dog should sit without being told to with the cue of "which hand?"
6. With a treat in your hand, stand in front of your dog.
7. With your dog looking at your hands, switch the treat from one hand to the other. Allow your dog to watch you make a fist to hide the treat.
8. Say "which hand?" Remember to make sure your hands are not close together so your dog can easily let you know which hand the treat is in.
9. Once your dog successfully choses the correct hand, place both hands behind your back and show your dog two fists. Place your fists close to your dog's nose.
10. Say "which hand?" If your dog chooses the correct hand, allow them to eat the treat. If they chose the wrong hand, open the hand with the treat in it. Don't say or do anything as you want your dog to notice the treat is in the other hand. Don't let her take the treat. Instead, close your hand to make a fist and ask your dog "which hand?" Repeat this until your dog picks the right hand.

Case Example: Prince Picks a Hand

Prince is a senior dog who used to be highly active. He took part in dog sports and even won a few awards. Unfortunately, over the last couple of years, Prince has struggled with arthritis and is losing his hearing. Because of this, Prince can't take part in a lot of the tricks he loved to do.

Since Prince can't do much, his companion, Denise, has looked at different tricks to teach her dog. She knows that one of the best ways to keep Prince happy is to keep his mind active.

Book 2 - Tips And Tricks To Dog Training

While Prince is laying beside the couch one day, Denise sits down by him with a treat. Prince watches as Denise places the treat in her right hand. She shows Prince the treat and then makes two fists. Knowing that Prince won't hear her well, she gently shakes her right hand. Prince bumps her right hand with his nose, causing Denise to open up her hand and allows Prince to eat the treat. Because of Prince's age and declining health, Denise only repeats this trick twice during a session. They have four training sessions a day.

Within the week, Denise places the treat in her hand before she sits down by Prince. While Prince does not see the treat, he knows one hand has the treat. He waits for Denise's little shake of a hand to pick which hand. At this time, Denise doesn't shake right away because she is trying to get Prince to smell the treat in her hand instead.

One day, during the second week of this training, Denise sits down and within a few seconds, Prince bumps his nose on her left hand. Smiling, Denise opens her hand to show Prince that he picked the correct hand. While Prince doesn't always get the correct hand, he is starting to understand he needs to sniff both hands and one will have a stronger treat smell than the other.

The story of Prince and Denise shows that it's perfectly fine to change the steps of a trick, as long as you don't lose sight of the point for the trick. Denise had the goal that Prince would pick the hand which held the treat through smell. Because he doesn't hear well, she knew that using verbal cues would not help. Therefore, she used a visual cue to help Prince learn the trick.

Chapter 7: So, You Want to Train Your Dog to Dance and Other Tricks

Keeping Your Dog Active

The reason people continue to train their dog, no matter what age, is because it helps keep their dog active. You want to keep your dog as active as possible, without causing them to become too tired or stressed, physically and mentally. There might be times where your dog can't do certain tricks anymore. If this happens, try to find a new trick to teach your dog that will be appropriate for their age and ability level.

Fortunately, for those long winter days or dogs that can't handle a lot of tricks, there are other ways to keep your dog healthy and active.

- Have them find treats around a room. Don't spread the treats everywhere unless your dog can handle spending a lot of time on their paws.
- Walk around the house with your dog.
- Play inside games, such as tug and release.
- Purchase toys that you can place food and treats inside. Give them to your dog and let them play and work for their treats.
- Hide your dog's toys and tell them to find their toys. You will want to give their toys names and teach them these names before you play this game.
- Practice tricks they know and can perform.

Always Remember Safety

One of the biggest reasons you will teach your dog basic tricks is because it will help keep them safe. For example, you can teach your dog to sit and stay if you are near a road, so they don't run after a car or a dog they find interesting. You will teach them "get down" so they don't hurt someone by jumping on them or don't climb up on something they can fall off.

You always want to keep in mind what your dog can and can't handle when you are training. If you try to train your dog to learn a trick and you find them struggling too much, you may need to focus on an easier trick.

With over a hundred tricks that your dog can learn, it's impossible to get them all in this book. However, I wanted to focus on different tricks because each dog breed will be different when it comes to learning tricks. Some dog breeds will strive on agility training while other dogs are more interested in hunting. If your get a dog from a shelter who is older, there are several tricks you can't teach your dog because of their health and age. But you can always teach dogs the basic tricks and then branch out into some of the easier tricks.

The tricks in this chapter are going to focus on basic, intermediate, and advanced tricks. Remember, you want to start basic, but how quickly your dog advances with the level of tricks they learn depends on their personality and your consistency, not their age.

Traditional Favorites

Sit Pretty

The trick to sit pretty is also known as beg. This is a trick a younger dog will do as it involves the dog sitting on their back paws and holding up their front paws in the air. Dogs who are older may have a problem with this trick because it will put too much pressure on the joints of their back legs, which is bad for arthritis.

The verbal cue for this trick is the word "beg." The visual cue is you placing both your hand up by your chest and holding your fingers down.

1. Command your dog to come to you. Praise them for listening to

the command.
2. Tell your dog to sit.
3. When they are sitting, use a treat to lure your dog's head up and back. At the same time, you want to say, "beg." Don't drop the treat into your dog's mouth or beside your dog. Let your dog nibble the treat from your fingertips. This will help your dog stay in the position longer.

   At this point, you don't want your dog's front paws to lift up off the floor. If they do, the treat is too high for them to reach and you should lower the treat for them.
4. Repeat this trick for a few sessions until you notice your dog holding better balance.
5. With the treat in your fingertips, repeat the same process, but raise the treat a little bit so your dog has to stand on their back legs. Again, let your dog nibble on the treat so they can work on their balance. If you have a larger dog, stand behind your dog and move the treat from the front of their face back toward you.

   Don't go so far that your dog decides to turn around. You want your dog to work on standing with their two back legs. It might help your larger dog to lean against you and you hold their chest to keep them in the position at first. Soon, they will be strong enough to handle the position themselves.
6. When your dog's balance starts to improve, move away and place your hands in the visual cue position and say "beg." Toss the treat to your dog. The key is to toss the treat when your dog is still standing in the right position.

Speak

If you are trying to train your dog not to bark but becoming frustrated because they don't listen, you may want to try this trick. Speak allows your dog to bark on cue instead of at whatever causes them to bark. The verbal cue for this trick is "speak" or "bark." The visual cue is holding your hand in the "L" shape.

1. Before you start training your dog to speak, you want to take time

to observe your dog. What makes your dog bark? Is it the mailman, a doorbell, or seeing you drive up the driveway? You will want to pick one factor that makes your dog bark as stimulus for the training. For the purpose of this trick, I am going to use the sound of a doorbell.

2. Standing in front of your door, say "bark" and ring the doorbell. When your dog barks, reward them with positive reinforcement. Repeat this step a few times before you move on to the next step, which needs to be in the same session.
3. Stop ringing the doorbell and tell your dog, "bark." If your dog barks, give them a treat. If they don't bark after a few times, go back to the previous step.
4. Once your dog barks when you tell them to, but not when they hear the doorbell, go into a different room. Tell your dog to "bark." If they are hesitant or don't bark after a few times, return to the front door to practice some more.

Shake Hands

When your dog learns to shake hands, they will raise their paw to their chest height. Anyone who wants to shake your dog's paw will need to from this height. It is important that you don't force your dog to perform this trick if they are showing signs of fear. Many dogs become fearful around strangers or in a room with a lot of people. You want your dog to shake hands with people they are comfortable with.

The verbal cue for this trick is "shake." If you want to teach your dog to shake hands with both paws, you will have "shake" for one paw and "paw" for the other paw. The visual cue to use is holding up to fingers. If you do this visual cue for a different trick, you can point the two fingers down or in a certain direction.

1. Command your dog to sit in front of you. Praise them for listening to your command.
2. With a treat hiding in your right hand, place it low to the ground so your dog can easily touch your hand.

3. Encourage your dog to paw at your hand by saying "get." When your dog lifts up their left paw, give them a treat. When your dog starts to raise their left paw when your hold out your hand, change "get" to "shake."
4. Lift up your hand a bit and say "shake." Once your dog lifts their left paw up to their chest, reward them.
5. Once your dog catches on to the verbal cue and raises their paw to their chest, start using the visual cue with the verbal cue. Eventually, you will stop using the verbal cue and stick with the visual cue.

Silly Dog Tricks

Act Ashamed

Act ashamed is an advanced dog trick that people love to teach their dog because it looks cute. Most people don't teach their dog this trick. However, many dogs start doing this when they do something you don't approve of because they are ashamed. Your dog is smart, and they will quickly catch onto the meaning of this trick.

Most people use the verbal cue of "shame." However, you can use a more positive word if you want to make sure this trick stays silly for your dog. The visual cue is often one finger held up.

1. Use a cushion that is affixed to the back of a chair. Show your dog that you have a treat in your hand. Place the treat under the cushion, but in the front so it is easy for your dog to get to. Tell your dog to "get" to encourage them to go for the treat.
2. Over time, you will place the treat further under the cushion, causing them to bury their heads into the cushion to look for the treat. After a few times, say "shame" when your dog buries their head under the cushion to look for the treat.
3. Slowly, you will stop using the treat and tell your dog "shame," which gives them the cue to look under the cushion for a treat.
4. From the back of the chair, reach into where your dog's mouth is

and place a treat near it, allowing them to grab the treat.
5. After a few times, hold the treat in your fist. When you notice them smelling your hand for their treat, tell them to "wait" and release the treat in a couple of seconds.

Expert Tricks

Find the Car Keys or Remote

It happens to everyone: We all lose track of our car keys. We often wish that we could tell our dog to help us look for our keys. Well, there is a trick to teach your dog to help find your car keys or remote.

If you train your dog to find both the car keys and remote, you will have two verbal cues. You will either say, "keys, find it" or "remove, find it."

1. Fill a small key chain bag with a few treats and attach it to your car keys.
2. Toss the keys in a playful manner and tell your dog "keys, fetch." When your dog brings the keys back to you, open the bag and give your dog one of the treats. Your dog will know there are treats in the bag before they bring it back to you. Because they can't open the bag without bringing it back to you, they will quickly learn to bring you the keys. If they do try to open the bag instead of brining you treats, this is usually the case when it comes to puppies, gently take the keys from them and try again.
3. After a few successful sessions, cue your dog by saying, "keys, find it," instead of "keys, fetch."
4. Once your dog constantly brings your keys back to you, hide them in a room.
5. Command your dog to come to you.
6. Say, "keys, find it," and help your dog search the room. Allow your dog to find the keys so they bring them to you. Reward your dog with a treat and praise.

Get the Mail

Most dogs love to have anything to do with the mail. Your dog may have a connection with your mailman, they may enjoy playing with the newspaper or have made a mess of your mail trying to grab it. You can teach your dog to bring the mail to you without ruining any of your bills.

The verbal cue for this trick can be different phrases as it depends on who the dog is bringing the mail to and where. For the purpose of this trick, I am going to use "bring to the table" because the dog is to place the mail on a little table by the door. This cue won't work if a table is too tall for your dog. You can always ask them to bring it to you, someone else, or train them to set it down somewhere.

1. Place an envelope, some mail, or a piece of paper where the mail drops at your door.
2. Stand by the table you want your dog to place the mail at with a few treats.
3. Command your dog to "come" and praise them when they do.
4. Walk your dog to the mail and tell them to "take it." To help them understand, you can point to the mail. Because this is more of an advanced trick, your dog may understand you with the phrase "take it" or your pointing. If they don't, place the mail into their mouth and help them set it on the table.
5. If your dog grabs the mail themselves, point to the table and say "drop" once your dog is close enough to the table. Reward your dog and repeat the process.
6. Every time the mail comes, meet your dog at the door, so you can give them their commands.
7. Once your dog understands to grab the mail, change the verbal cue to "bring to table" instead of "drop." Over time, your dog will automatically bring the mail to the table and you won't have to use your verbal cues. However, you will want to ensure that your dog is doing their job. If they need reminders, use the verbal cues and give them praise when they follow through.

Dancing

You should only teach any type of dancing or sports tricks to dogs who are really active. If you have an older dog, you can teach them some dance moves, but you will want to make sure they have a strong grasp of training and repetition won't bother their joints or cause them pain.

Heel Forward and Backward

The two cues for this trick are "heel" and "back." The visual cue is to make a fist and place it on the side of your stomach. The dog will always walk on your left side when they are doing heel walks. Your dog should also sit when you stop.

1. To help guide your dog, you may want to place them on a loose leash and set them on your left side.
2. Say "heel" and walk forward with your left foot first. Your left foot will soon become your dog's signal to heel.
3. Each time your dog completes a heel command, praise them with positive reinforcement.
4. When you stop, start to slow your walking. Place your left foot flat on the floor. With your right foot, hold it up in the air and set it right beside your left foot. Command your dog to sit. Reward your dog with positive reinforcement once they sit. The cue for your dog to sit will become your right leg going up and then landing next to your left foot. Your dog will also learn that this cue is soon coming because you slow down your walking pace.
5. To have your dog walk back, say "back" and take a step back. If your dog doesn't motion, move them with the leash or guide them to take a step back. Once they move, reward them with positive reinforcement.

Take a Bow

Teaching your dog to take a bow is relatively easy and can be done during the basic training stage. The verbal cue is "bow" and your visual cue is placing one foot behind your other foot. The foot that is behind you should rest on tiptoes, just as your feet would do if you bow or curtsy.

When your dog bows, the position they stand in is keeping their legs upright and placing their front leg's elbows on the floor.

1. Command your dog to come to you. Praise them for listening.
2. With your dog facing you, have them stand.
3. Place a treat in your fist and hold it at the nose.
4. Gently, place your hand on your dog's nose and press downward. Give the cue of "bow."
5. When your dog's elbows touch the floor, release the treat.

If your dog doesn't bow down easily, you may need to break the trick down in steps. For example, give your dog the treat when they start to bend their front legs or place their head down.

Chapter 8: Common Mistakes

It's going to happen no matter how hard you try to make your dog's training journey as pleasant as possible—you will make a mistake. You will make several mistakes. However, you and your dog will often make mistakes together. When you start to train your dog, it is a lifelong commitment for your dog, and you need to be as consistent as possible for your dog to thrive through their training journey. It doesn't matter what trick they are learning or what they already know; consistency is one of the biggest keys to successful training.

However, you should never get frustrated when you or your dog makes a mistake in training. You always want to have the mindset that mistakes happen. Learn from your mistake and move on. Don't dwell on your mistakes and don't react harshly toward your dog when they make a mistake. Always remember that your dog is learning just as you are.

One way to keep yourself from limiting the number of mistakes made in the training journey is learning about the common mistakes and keeping them in mind. When you notice that you are on your way to making a mistake, you will be mindful enough to recognize it and stop yourself from making this mistake. If you do make a mistake, you are mindful enough to realize it, learn from it, and move on.

Mistake #1: Training Your Dog for Too Long

You always want to keep in mind that your dog can only train for a specific period of time before it starts to become boring, too stressful, or they become too distracted.

Dogs, no matter what age, don't have a long attention span and they can't focus on one task for several minutes at a time. A puppy's training session should only be a couple of minutes. Of course, obedience classes will go for a longer period of time, but this doesn't mean you should do this at home.

Obedience classes are run in a certain way by professional dog trainers. Allow them to handle the longer training sessions. You can start to train an adolescent dog for about a minute or two longer. An adult dog has an attention span that can last up to six or seven minutes. Senior dogs, unless they have health issues, can last about eight to ten minutes.

Just because you need to limit the amount of time, doesn't mean you can only train once a day. You can hold several training sessions throughout the day for your younger dogs. Your older dogs, especially senior dogs, should be limited to what they can handle. You will learn what your dog can and can't handle by observing your dog.

Some people will use a timer to help them stay within the training limit they set for their dog. People also schedule training sessions every couple of hours for younger dogs and two to three times a day for older dogs.

Mistake #2: You Lack Consistency

Remaining consistent is the main problem people have when it comes to training. However, it is a key when it comes to training your dog in any trick. Even when your dog comes to learn the trick confidently and tends to follow the trick automatically, you still want to make sure that you don't miss a step or don't forget to give them the cues you need to.

One of the hardest parts is correcting your dog's behavior when it comes to training. Many people will use the e-collar to correct their dog's behavior through tricks. However, if their dog does something they aren't supposed to do and the owner doesn't have their remote to give their dog a signal handy, they are going to miss the opportunity. If you decide to keep your dog's e-collar on throughout the day (they should never wear it when they are sleeping), you want to keep the remote with you throughout the day. You can always purchase a smaller remote that can clip on your belt or fit easily into your pocket.

Book 2 - Tips And Tricks To Dog Training

You always need to remember, whether it comes to correcting your dog's behavior or rewarding the behavior, it has to occur within a few seconds from the behavior. For example, if you are teaching your dog not to dig in the garbage behind the garage with the e-collar, you want to send your dog the signals while they are digging. If you do this after or before they can associate the signal to what they are seeing or doing at that time. If you want to praise your dog because they picked up their toys without you commanding them to, you want to do this right after they pick up their toys or they won't understand why they are being praised.

Mistake #3: You Wait Too Long to Start Training

While you don't want to start training your dog the second they walk through your dog, you do want to start within the first week or two, no matter how old your dog is. Even if you brought home a senior dog, start training them the basics. If they know the basics, they will quickly catch onto your cues or simply act on the trick when you give the cue. This often happens when it comes to "sit" because this is the common cue for the command.

You also want to remember that the same day you bring your dog home, you will start training them, it is just not through sessions. For example, you will show them where they will eat, use the bathroom, and sleep. Always remember, whenever you teach your dog something new, it is training.

Mistake #4: You Use Harsh Discipline

Training is a form of discipline as it helps your dog understand how they should act and often corrects their behavior. There is no need to ever hit, kick, or even place your dog in a time out for their behavior. Instead, you want to find a way to train your dog to replace the unwanted behavior with wanted behavior.

This is something that will take time, but your dog is going to trust and respect you more than if you harmed your dog through hitting. Your dog will not become fearful of you, they will learn to listen to your commands and want to do so because they want to make you proud.

Mistake #5: You Stop Training

You should never stop training your dog. Even when your dog is older, you should always practice the tricks they can still complete. For example, if your dog has arthritis, you won't want to focus on agility training, hunting, or dancing. However, they can still sit and lay down. Keep your eye on these tricks as they will help your dog remain active and keep their mind going at the same time.

Your dog will let you know when a trick is becoming too much for their body. You always need to pay attention to your dog's body cues and follow them thoroughly. Your veterinarian and professional dog trainer can also help you understand if you can't read your dog well.

Mistake #6: You Don't Train in the Right Mindset

Training takes a certain mindset. You need to be calm and happy to train your dog, even in the basic training level. Always remember that your dog will feed off your emotions. This means when you are angry, your dog will feel this and cause them to react in fear or become aggressive.

One of the biggest ways to make a dog aggressive is by becoming aggressive toward the dog, including your tone of voice. Dogs are extremely sensitive to voice tones and will quickly pick up on a tone that is not happy or calm.

If you find yourself becoming frustrated or angry during training, or you feel this way before the training session, don't train! It's okay to skip a session or wait until your mood is lighter. Doing this will help the training session go easier and your dog will feel better when they are training.

Mistake #7: Abusing the Training Tools

If you use an e-collar or clicker, be sure that you don't abuse these tools. Some people feel both of these tools need to be used whenever your dog does something you don't want them to do. This is a myth. If you use these tools when you are not training, you will confuse your dog, and this will cause more problems when it comes to training.

More importantly, when it comes to the e-collar, you can harm your dog by sending them signals too often. You won't hurt them physically, unless you don't move the collar around every couple of hours or the collar is on too tightly. You will harm them emotionally and mentally.

Conclusion

There is a lot of information within this book, and while you might feel a little overloaded, as many people do when they read a dog training book, you also feel more confident about training your new family member. Whether you brought home a puppy or an older dog, you have a great amount of information at your fingertips to help you and your dog through the training journey.

You can use this book to help your dog learn tricks from the puppy stage all the way to becoming an adult. At the same time, you can remind yourself of the common mistakes people make so you don't catch yourself doing the same thing. Remember, remaining mindful of your training, supervising your dog, and knowing where any of your training tools will help you stay consistent.

Some of the key takeaways from the contents of this book that you should always remember are:

- **Consistency**: You always need to be consistent as a dog trainer. Even if you aren't in the mood to get up and praise your dog for exhibiting great behavior, you need to do this. This means it is always helpful to have tools such as treats next to you so you can quickly grab them as a form of positive reinforcement. At the same time, you always need to ensure that you are correcting your dog's unwanted behavior through tricks.
- **Always use positive reinforcement:** One of the biggest reasons training works so well is because people use positive reinforcement when their dog performs a trick. Some people will even use it when their dog tries but fails or makes a mistake to ensure they keep trying. Dogs thrive on positive reinforcement. They love getting attention from you, so when you are training them, make sure they

get as much attention as possible. Positive reinforcement is more than giving your dog a treat!

- **Never use harsh discipline:** Training your dog is a form of correcting their unwanted behavior and transforming it into wanted behavior. As long as you consistently train your dog and help them through the training process, you will never need to discipline your dog. Instead, you will find a trick that will help your dog learn what to do and what not to do. This means you should never hit, kick, or knee your dog. You never need to perform any type of action that is going to make your dog fear you in any way. The stronger relationship you and your dog have, the better training will go for both of you.
- **There is never a perfect time to start training but start after your dog is comfortable:** You never want to start demanding your dog to do or act a certain way the minute you walk them through the front door. If they do something they shouldn't, distract them with a toy, but don't give them positive reinforcement. You want your dog to become comfortable with their new environment and the people within it. This includes other pets.
- **Always observe your dog's behavior:** Your dog will tell you when training is going on for too long. They will tell you when they are not feeling well or are too hot. They will tell you when something is wrong. Even if you are in the middle of training, you have to stop and pay attention to their behavior. If your dog is tired, let them rest. It is more important to keep your dog healthy than making sure they are finishing a certain task.
- **Always teach your dog gradually and with small steps:** The steps written within the contents of this book are not written in stone. You can break apart the steps to help your dog learn the best way for them. Realize that training is a lifelong commitment for your dog. Once you start, you will train them until they can't take part in the activities anymore. This means you have plenty of time to teach them all the tricks you want to. Go slow and let your dog set the pace. Some dogs will learn faster than other dogs and

this is fine. It doesn't mean something is wrong with your dog, it simply means they learn at a slower pace than your friend's dog.

- **Have patience:** Patience in training is just as important as consistency. You want to train when you are patient and stop when you feel like you are losing your patience.
- **Never train when you are angry:** Training when you are angry is one of the top ways to teach a dog to become aggressive. Your dog will associate the feelings they get from you to their training. Furthermore, if you are angry your dog might think that they did something wrong when you are angry about a different situation that is completely unrelated to your dog and training. Even if it is time for your dog's at-home training session, don't train if you are angry. You want to wait until you are happy and excited about training as this feeling will be the way your dog feels about training.
- **Your dog's mental and emotional health is just as important as their physical health:** It is usually easier to understand when something is wrong with your dog physically. It is harder to know if your dog's emotional and mental health is not at a good place, especially if you just brought them home. If you brought home an older dog from the shelter, they could have had traumatic experiences that are still haunting them. Take them to the veterinarian and even a dog behaviorist to make sure everything is good to go before you start training.
- **Dental hygiene is important:** One factor that many people don't think of is their dog's dental hygiene. However, this is just as important as their physical health. Get treats that will help your dog brush their teeth if they don't allow you to brush their teeth. Make sure your dog has dental check-ups just like they do with physical check-ups.
- **Have fun!** Even if your dog is learning agility training for dog shows and training is very serious and demanding, you want to make sure that your dog has fun with training. You always want to include play time and make sure they have a period of rest and get enough water, especially on a hot day. Following these tips will help your dog have the best training experience possible.

Book 2 - Tips And Tricks To Dog Training

References

5 Tips for Dog Training Session Prep - Acme Canine. (2017). Retrieved from https://acmecanine.com/5-tips-dog-training-session-prep/

10 Best Training Tips. Retrieved from https://www.pedigree.com/dog-care/training/10-best-training-tips

Becker, M. (2014). The 3 Most Common Dog Training Problems and How to Fix Them. Retrieved from http://www.vetstreet.com/our-pet-experts/the-3-most-common-dog-training-problems-and-how-to-fix-them

Bender, A. (2019). Top 5 Ways to Use Positive Reinforcement to Reward a Dog. Retrieved from https://www.thesprucepets.com/ways-to-reward-a-dog-1118276

Bolluyt, J. (2018). 18 of the Easiest Dog Breeds to Train. Retrieved from https://www.cheatsheet.com/culture/easiest-dog-breeds-to-train.html/

Clark, M. 7 Most Popular Dog Training Methods - Dogtime. Retrieved from https://dogtime.com/reference/dog-training/50743-7-popular-dog-training-methods

Conn, K. (2015). 5 Steps To Correct Inappropriate Dog Chewing. Retrieved from https://www.cesarsway.com/5-steps-to-correct-inappropriate-dog-chewing/

Dog Ages & Stages - Dogtime. Retrieved from https://dogtime.com/dog-health/dog-ages-and-dog-stages/253-ages-stages

Gibeault, S. (2018). 13 of the Most Trainable Breeds. Retrieved from https://www.akc.org/expert-advice/lifestyle/13-of-the-most-trainable-breeds/

Huston, L. Caring for Senior Dogs. Retrieved from https://www.petmd.com/dog/care/evr_dg_caring_for_older_dogs_with_health_problems#

Johnson, D. (2013). Ten Tips for Taking Care of Your Dog. Retrieved from https://iditarod.com/ten-tips-for-taking-care-of-your-dog/

Klinger, C. (2019). When Should I Start Training My Puppy or Kitten? | Hill's Pet. Retrieved from https://www.hillspet.com/pet-care/training/when-to-start-training-puppy-kitten

Lotz, K. Top 10 Senior Dog Training Tips. Retrieved from https://iheartdogs.com/top-10-senior-dog-training-tips/

Obedience Courses. Retrieved from https://www.animalhumanesociety.org/behavior/obedience-courses

Relationship-Based Dog Training: Benefits. Retrieved from https://resources.bestfriends.org/article/relationship-based-dog-training-benefits

Rollins, J. (2016). How to Discipline a Puppy or Dog: Effectively Punishing Your Dog. Retrieved from https://www.petexpertise.com/dog-training-article-using-physical-punishments/

Stages Of Puppy Development. Retrieved from https://dogtime.com/puppies/1130-puppy-behavior-basics-hsus#/slide/1

Tucker, N. (2018). Adolescent Dogs: 6 Facts To Know - Whole Dog Journal. Retrieved from https://www.whole-dog-journal.com/puppies/adolescent-dogs-6-facts-to-know/

BOOK 3

TRAINING YOUR PUPPY STEP-BY-STEP

A How-To Guide to Early and Positively Train Your Dog. Tips and Tricks and Effective Techniques for Different Kinds of Dogs

PAUL DAVIS

Book 3 - Training Your Puppy Step-By-Step

Introduction

One of the greatest joys in life can be the little pets that we have at home. Watching a dog grow from a teen puppy to a well-trained furry companion is an exciting thing. Many individuals believe that having a puppy is going to be a fun experience. It certainly is! At the same time, the hard parts of training are often overlooked. Some people can get in over their heads and end up with too much to handle when trying to properly train their four-legged friends.

You might be the most responsible pet owner out there. You could have only fresh organic treats and food, the top-quality grooming and veterinary care, and a personal trainer for your dog. No matter what you do, there are no circumstances that will keep you free from the slight chance that you might wake up in the morning to find a trail of diarrhea across the house, an expensive shoe that's been ripped up, bits of a dollar bill found in vomit, a school paper soaked in pee or a wad of toilet paper dug up from the trash can and ripped to bits strewn about a house.

Dogs are cute, but they are animals, which means they aren't free from having their nightmarishly disgusting moments. They might get fleas, ticks, and other creepy critters that can make your stomach turn. This isn't to turn you off from getting a dog! Nor are these aren't daily experiences. However, they are occurrences often overlooked by those wanting to adopt a pet, which is why millions of household pets end up in shelters every year. Don't think that having a dog is all disgusting, but don't think that you'll be free from these random instances.

Nothing we discuss is meant to scare you away or make you nervous about having a dog! With all the animals on the streets, in shelters or even in abusive and neglected homes, it is important to know everything that you are getting yourself into before adopting a pet. These are innocent animals who are going to look up to you from the moment you bring them home.

Don't think of your puppy as a human. If anything, think of it like a human baby, but throughout its entire life! They will be smart enough to learn new tricks, but they will also not realize that eating garbage is wrong. Throughout this book, we will give you a comprehensive guide of everything you need to know so you can successfully raise a healthy pup.

There are 8 steps to training your puppy:

1. Pick out your dog.
2. Prepare your home.
3. Start basic training.
4. Socialize them and prepare them for the world.
5. Teach them more challenging skills.
6. Ensure there is proper exercise.
7. Keep up with the health of your dog.
8. Avoid the mistakes and reflect on what you can improve.

Throughout each chapter, we will give you all the info necessary to help you create your training plan. In the final chapter, we will discuss how you can create a specific schedule along with other important reminders to live by throughout this process.

It is an exciting time and you can be assured once you bring your furry friend home, life will never be the same! Let's get started now in discussing how to choose the right dog for your home and your lifestyle.

Chapter 1: Choosing the Right Dog

Most people give up their animals because they can't afford them (Greenwood, 2016). At the same time, many individuals will end up giving their dogs away because they simply can't keep up with their basic needs.

Picking the right dog is important. There are a few options that you have to get started. You could go to a shelter, a breeder or a pet store. Once you have decided where to get your dog from, do your research on the specific breed. Throughout this chapter, we will give you tips on how to best choose the right dog, along with a few suggestions for people with specific needs. If you don't get a dog that matches your lifestyle, it can make training them more challenging, while weakening the relationship. Consider all factors that we discuss in this chapter when choosing the new family member you will be adding to your life.

Picking Where Your Dog Comes From

With over 900 million estimated dogs in the world, we can be certain there won't be a shortage anytime soon (World Atlas, n.d.). When choosing to bring a new dog into your home, you'll have several different options to choose from. There are a few important things to remember when picking the right place.

First, you want to ensure that you are getting a dog from a place you support. Before adopting a dog from anywhere, whether it be a puppy store, a breeder or a shelter, do some background research as to how they train their dog. Of course, the dogs aren't at fault for how they are treated, but it is best to make sure that we are supportive of those who care for their animals properly.

It is also important to remember that puppies will always be in higher demand, but do not overlook older dogs! We will be focusing on training your puppy throughout the book, but don't forget that contrary to the popular saying, old dogs can learn new tricks too! Even if you adopt a dog at 10 years old that isn't potty-trained, has a problem with barking, and has almost no structure, you can still train them. The difference between an older dog and a puppy in training is that puppies are simply easier to train.

When deciding on *where* you will be getting your dog, keep in mind that every year, over two million dogs are given to shelters. Not even half of them will end up adopted (ASPCA, n.d.). There are plenty of puppies in shelters as well. Just because a dog isn't the size of a toy animal doesn't mean it is not a puppy! A puppy is any dog less than a year old. Between their first and second years is when they are considered adults, but even then they won't always act like an adult.

After about six months and up until eight months, dogs grow out of their

Book 3 - Training Your Puppy Step-By-Step

'puppy' looks as well. They get bigger, their ears and face grow into an adult's, and their coats are usually fully grown in by this point. Though puppies are adorable, we will only have them as puppies for six months, so it is essential that we make the right decision for our future, as well as our dog's. You have several options from where you can choose a dog. These are the most common places that people get dogs, according to AnimalSheltering.org:

- Adopted from Shelter (44%)
- Bought from Breeder (25%)
- Pet store (4%) (Animal Sheltering, n.d.)

The remaining were smaller and random statistics, such as adopting from a friend, received as a gift from a relative or taken in as a stray.

Let's start by understanding what a shelter is. The first kind are municipal shelters. These are run by government officials and the animals are usually pick up off the street. Compassionate employees ensure population of certain dogs and cats are under control. They also ensure that sick and injured animals are given the treatment needed. These shelters are often overrun, so as a method of controlling population, animals will be euthanized after certain periods of time, or under special conditions in these shelters. Adopting a dog from a municipal shelter might be the best method to save a life!

There are private shelters as well that will take in animals under some of the same conditions. They will also adopt animals from caretakers that can no longer handle them, and they might even partner with municipal facilities. Private shelters might euthanize, but many strive to be no-kill as well. It is up to you to discover private shelters in your area and do your research to see what their adoption policies might be.

Humane societies and SPCAs will also have animals up for adoption if you have one in your area. These are more commonly known for their advocacy for animal rights, promoting education or aggressive neutering and spaying efforts to keep population down. Some are no-kill and others aren't. It is important to remember that even in no-kill shelters, overcrowding can occur, meaning that sick animals might not get the treatment they deserve.

Breeders are usually smaller family homes that make it their job to breed the dogs they have to sell the puppies in a new litter. Breeders will usually have one specific breed, giving you the option to seek out a type of dog specifically. Breeders are much more regulated than they used to be, and many veterinarians will have recommendations for you to find one which is more reputable. If adopting from a breeder, make it your mission to meet them and the puppy before adopting.

Sometimes you might have to travel far to do so, but it is worth the trip. Sometimes, puppies are kept far from any humans and live in shelters or sheds away from the home of the breeder. This is a sign that the animal might not have had enough socialization. Of course, we want to help all dogs, especially those in bad conditions. However, if you have a family of five children who are still young, an anti-social dog isn't going to be good for you. The best way to ensure that your new furry companion becomes a lifelong part of your family is to make sure it fits in with you and your life in the first place.

A final option you have is through a pet store. These are very controversial because of the use of puppy mills in order to populate the store. The Humane Society of the United States conducted investigations under cover and made disturbing discoveries while using hidden cameras. They reported that there were "significant violations, including sick and injured dogs who had not been treated by a vet, underweight dogs, puppies with their feet falling through the wire floors, puppies with severe eye deformities, piles of feces and food contaminated by mold and insects (The Humane Society of the United States, n.d.)."

The first thought is that these puppies should be purchased to help keep them from having to live in these conditions! Unfortunately, what happens is that this increases demand, meaning that more puppies would have to go through this. Puppies that aren't purchased from the store will end up being given away if they haven't been purchased after a long time. You are entirely in charge of your own decision to purchase from a pet store if you'd like, but as a responsible pet owner, you should be doing proper research first.

This is a big decision for you to make, which affects both your personal and family life. There are many benefits and disadvantages of certain places, and the welfare of dogs is highly sensitive in many societies. For this reason, certain places might be labeled as immoral, but at the end of the day, it will be up to you to decide!

Energy Levels of Certain Dogs

Now that you have a good understanding of where to get your dog from, let's talk about factors that you will need to consider when picking your breed! The most important part for picking your breed is understanding their energy levels. This is because it will directly affect the kind of care they need. Some breeds might need a huge yard that they can run in on a daily basis. If they don't get this time outside, it could increase their aggression or make it difficult to control them when they are excited. Other animals might only need a couple short walks a day and are perfectly fine in social settings. To make sure that you are picking the right animal, you'll want to consider how energetic they are.

Individuals who work long hours with jobs that take up a majority of their day-time should have lower energy dogs. If you can't take your dog for walks during the day and only have time for a quick potty break at night, a large, high-energy dog shouldn't be your first pick. If you have children who can help care for the dogs and who will be playing with the dogs frequently, a high-energy dog is perfectly fine. Single individuals who live alone would probably want a low to moderate energy dog. If you don't have access to a yard and live in a smaller apartment, a low energy dog is first choice. Having access to a dog park or a regular park and beach is great for your dog and can help provide them with a lot of freedom and energy.

Consider their energy levels for when you are training them as well. Some dogs will be able to train easily and will know right away what you are commanding them to do. Other dogs might be harder to control because they are so playful and active. Being active or not isn't something that is good or bad. It will all simply depend on your own personal energy level and how much you'll be able to contribute to the dog training process.

Lower energy dogs are those that are considered toy breed dogs, and the non-sporting group of dogs. Toy dogs have the lowest energy. However, this doesn't mean they aren't necessarily active. For example, a Pomeranian has low energy because you won't need a large yard for it to run, but it could still be the loudest barker in your apartment complex! Toy dogs were bred to be a companion to humans. They are specifically used to be carried around and sit on the laps of their owners. They will be best for those who are looking for a dog to have as a friend. These are best for older individuals, those with mental illness that might want an emotional support animal, or single individuals that aren't as active. If you are single or live with a partner or perhaps you live in an apartment complex, this is the best category for you. It doesn't mean that your dog just lays around all day either! All puppies are very active and will require training to prevent barking, jumping, and other high-energy characteristics. However, as time goes on, low-energy dogs will be the kind that become more relaxed and require less physical activity than medium and high energy dogs. These are breeds typically known for having low energy:

- Bichon Frise
- Boston Terrier
- Bulldog
- Basset Hound
- Bloodhound
- Mastiff
- Manchester Terrier

- Pekingese
- Papillion
- Irish Wolfhound
- Chow Chow
- Affenpinscher

High energy dogs are those that stem from the terrier, sporting, working, and herding groups. These are dogs that were bred to be by your side at all times. Hunting dogs were trained to retrieve the flying animals (duck, etc.) that their owners shot down. Herding dogs were trained to help round up farm animals such as cattle or sheep. Herding dogs love having wide ranges to run around, and they will certainly try to naturally herd humans, especially children.

Working dogs are those that have been trained for protection. They might also pull things like sleds or be trained for rescues. Terriers were frequently trained in order to help protect the home and hunt smaller animals. Dogs

Book 3 - Training Your Puppy Step-By-Step

from these categories will want to swim, play games, and spend a lot of time outdoors. They are very friendly and intelligent as well. If you are someone who loves spending time outdoors, has a huge yard, wants a dog for a specific task (hunting, fishing, farming, etc.) then you will want a high energy dog.

If you want a dog with high energy, these are dogs you should consider:

- Australian Shepherd
- Shetland Sheepdog
- Russell Terrier
- Siberian Husky
- Dalmatian
- Border Collie
- Labrador
- Golden Retriever
- Beagle
- Poodle
- German Shepherd

Medium energy dogs are those from all kinds of breeds but are known for having a nice balance between being able to play outside with families, but also be fine with sitting on the couch with you while you binge watch TV. This is perfect for those who have a fairly active lifestyle but aren't quite the kind of person who spends the majority of their time outside. These are also the dogs that are perfect for families who want to include a pet but might not have as much time for their animals, like a single individual or a young couple might have. For medium-energy dogs, you might consider:

- Australian Terrier
- Alaskan Malamute
- Akita
- Bernese Mountain Dog
- Bergamasco
- Bloodhound

- Corgi
- Chihuahua
- Cavalier King Charles Spaniel
- Cocker Spaniel
- Dachshund
- French Bulldog
- Maltese
- Many Terrier breeds

Consider all aspects of your life and the way that you do things now. Getting an active dog doesn't mean that you will become active yourself necessarily. Dogs can do a great job at teaching responsibility to different owners, but it is best you get a dog that fits in with your lifestyle rather than trying to fit in with the lifestyle of the dog you choose.

Child-Friendly or Not

Having a pet for your family is a great addition. It is another member who you can share memories and joy with. It can teach you and your children responsibility, compassion, and kindness. Some of the best memories that children have include their pet! When considering a pet for your family, there are a few things that you will have to remember.

Consider the age of your children. If you have teens, then you can assume they will interact with a dog similarly to adults. Children between 10-14 might not be as active, so they can still be fine with a family dog that works for you as parents. If your child is younger or petite in size, no matter what their age, consider this if you are adopting a larger dog, as they might still end up being on the herding end of things!

Book 3 - Training Your Puppy Step-By-Step

For any family that has a young and active child, especially more than one, then you have to be very careful when picking out a dog. Dogs are sweet, kind, and loving. However, they also have the ability to attack and kill. Dogs don't need to be feared, but we have to remember that just like humans, they can act aggressively when provoked. Although you could train your dog and your child to interact peacefully with each other, when your back is turned, you never know how a dog might react if the child tugs on its tail, flails its arms, screams or does something else that could trigger an otherwise peaceful animal.

Dogs will act on instinct, and just like children, they don't always understand the right social cues. At the same time, remember that this is a fact for ALL dogs, so getting a child-friendly dog doesn't mean that you don't have to have a discussion with your kids about the proper way to care for it. Even a child staring at a dog for too long can be enough to give it mixed signals! These are the types of dogs that don't work well in households with children:

- Shih Tzu
- Siberian Husky

- Greyhound
- Pekingese
- Jack Russel Terrier
- Chihuahua
- Australian Shepherd
- Rottweiler
- Saint Bernard
- Akita

These dogs are less likely to have issues when you adopt them as puppies and raise them with your family. Again, this isn't to scare you away either. Most of these dogs are fine, and with proper care and training provided to both humans and dogs, issues can be avoided. These are simply dogs that are more likely to play rough, nip and bite, and growl or scare dogs. Many herding dogs will nip at the feet or ankles of those running around, so it is not always that your child is at risk for getting seriously injured, but they might learn to be fearful of the dog. Pick one that is right for the whole family and give you and your kids the chance to meet your dog and play with it before adopting.

Interacting with Other Animals

If you are adopting a dog, consider your other animals as well. We will discuss breeds that work best with different dogs but remember that some shelters might not let you adopt if you already have a certain number of animals. If you do plan to grow your furry family, ensure that all the pets you already have are up to date on their shots and you have documentation because there's a good chance you will be asked to provide proof of their medical history.

Cats and dogs have long been believed to be enemies. Part of this is because they are such different animals. Cats are perceived to be devious, sneaky, quiet, relaxed, and reserved. Dogs have always been perceived to be more active, excited, and friendly. One isn't necessarily better than the other, though this is a hot debate. They are entirely different animals with various needs and qualities that keep them as separate species. At the same time, having one or more of each in your home is a great addition to any family. Though they can be enemies, they can also be best friends.

The reason that dogs and cats might fight is because of their natural instincts. Cats are naturally more afraid of things that are bigger than them. They also need personal space and time to get to know new animals. If you walk into a room and pick up the cat sleeping on the couch immediately, it will not be so happy about this. If you walk into a room and start petting a dog, they're more likely to start wagging their tails and feel excited about your presence. Cats struggle with dogs because they aren't given that space!

Dogs struggle with cats because they are smaller and quicker, much like the rodents that they have been trained to attack. A cat might move across your house the same way an opossum or a raccoon might. Though they aren't rodents, they are still animals that move around just like the ones that dogs have been bred to capture.

Typically, it will be easier for you to bring home a dog and introduce it to a cat rather than the other way around. The cat might be angry at first, but after some hissing and hiding, maybe a swipe or two at the dog, it'll warm up. If you bring home a two-pound kitten and introduce it to your 80-pound Tibetan Mastiff, you might struggle to train the dog to be dominant whereas a cat would rather hide than try to fight new visitors.

Cats don't respect territory as much as dogs do either. Most cats are independent animals whereas dogs are used to being in packs. Of course, animals like lions are frequently found in families, but your household cat is very different from a pride of lions than it is from some dogs! Cats will walk up to you and sit on your lap whenever they want. Dogs are more territorial and when they see this, they might be likely to try and protect you from the cat. At the same time, you can have both in your home! The best kinds of dogs to bring home if you already have a cat are toy dogs and non-sporting dogs. This is because they will be more likely to listen to you, their pack leader, and trust that this animal isn't a threat. Obedient dogs that are well-trained will typically do fine with other animals anyway, but still consider your cat before bringing home another pet. Listed below are the dogs that will do the best with other animals, and even cats:

- Maltese
- Boxer
- Bichon Frise
- Cocker Spaniel
- Basset Hound
- Labrador and Golden Retrievers

If you are considering having multiple dogs, you will want to consider different breeds that work well together, and those that don't. Many households will simply get dogs that are from the same breed, and this is the safe choice. We should still consider those breeds that don't quite mingle the best together.

Book 3 - Training Your Puppy Step-By-Step

First consider your dog's past and if they have gone through anything negative with another dog. For example, let's say that you already have a mini poodle at home, and you want to introduce another dog. There is a pug that lives in the house next door that you think is cute and you'd consider buying. However, the pug barks and growls at your mini poodle every time you go outside together. Though the pug might not have anything against either of you personally and is just being territorial, your mini poodle will still have a negative perspective about pugs. They will associate their ear shape, face shape, and body shape with something scary, so don't bring anything home that looks like a pug in this case.

Some dogs are more territorial and are the 'alpha' dogs. If you have a strong territorial male dog already, do not bring home another male dog. You would instead consider introducing a female since they are less likely to be aggressive. For multiple dog homes, you will want to consider males more often as singular pets or only have one male with other females, and have multiple dogs be female. This isn't the case every time. there are plenty of households with multiple male dogs that do just fine. In any case, if you are picking out new dogs to introduce to the ones you already have, don't overlook this point.

Small dogs will tend to be a little more aggressive toward big dogs. They will feel threatened and as such will show their dominance to the big dogs since they feel as though their size might be lacking. If you have a Pomeranian at home and want to bring home a Chow Chow, then you just might end up finding that the Pom doesn't stop barking and continues to try and exert its 'alpha' stance.

Consider dogs with matching needs over all else. You don't want to get a Pekingese, who could sleep up to 2o hours a day, and then bring home a Labrador retriever who will need hours of play. If your lifestyle is this diverse, it could work, but you want your pets to be on similar schedules, so it is best to consider that their needs will match each other.

If you only plan on having one pet, and one dog, that is perfectly fine! There are many dogs that just want to hang out with you and you only. The dogs that prefer to be alone the most are:

- Great Danes

- Collies and Sheepdogs
- Poodle
- Pug
- Chihuahua

Grooming Needs

Some dogs are fine with a quick rinse and a brush, but others need bi-weekly or monthly grooming. If you get a dog that needs frequent grooming, but you don't have any time to groom them, they can get knots, matted hair, and start getting stinky much faster. Of course, many individuals like the look of groomed dogs and wish to have one for aesthetic purposes. Grooming can be expensive and tedious, so you'll want to consider the different grooming needs of various breeds.

The dogs that need the most grooming can also be the ones that are best for certain families, however. Hypoallergenic dogs are those that don't shed as much, so they are best for people with allergies. At the same time, since they don't shed, their hair will get rather long. If you don't properly groom them, their fur can get knotted. Ungroomed hair can collect sweat, dirt, and bugs. This could harm your dog's fur and make it easier to spread pests inside your home. If a dog has a knot, they can still be bothered by it, so they might chew and scratch, causing a more serious injury than the initial knot.

On the other hand, if you have dogs that don't need regular grooming, at least from a professional, they will still need some form of care and maintenance. All dogs will still need an occasional bath, and it is important to keep up with their nail and tooth hygiene. We will cover that in a later chapter. Dogs that are non-hypoallergenic shed their fur rapidly because they are constantly growing new hair. This means you will have dog hair on your clothes, furniture, and everything in between.

These are the dogs which will require the most grooming:

- Poodles – all variations (mini poodles, cockapoos, etc.)
- Bichon Frise
- Shih Tzu
- Maltese
- Yorkshire Terrier
- Mini Schnauzer
- Lagotto Romagnolo
- Irish Water Spaniel
- Cairn Terrier
- West Highland Terrier
- Scottish Terrier

These are the dogs which won't require frequent haircuts, but will shed a lot:

- Corgi
- German Shepherd
- Australian Shepherd
- Labrador Retriever
- Siberian Husky

These are dog breeds which will frequently shed and also require grooming and haircuts. These are not recommended for those with allergies or those who don't want to keep up with grooming:

- Akita
- Chow Chow
- Collie
- Pomeranian
- Pekingese

If you want to get a dog that requires grooming, of course, you don't have to know how to do this yourself. It is quick and easy to drop your dog off at the groomer's or pet shops that provide grooming services. This can actually help socialize your dog when done frequently. The biggest downfall of this is that you need to find the right groomer, which can sometimes be expensive. The more work your dog requires, the pricier it will be. If you are already struggling financially, don't get a dog that will need regular haircuts.

Other Breed Specific Conditions to Remember

Many of us have an idea of a breed that we want years before we even get a dog! All dogs are special, and they each have their own personality traits you'll want to consider. There are a few other factors you have to remember when picking a dog. Some people will give dogs to shelters over simple things, so to ensure you keep your dog around for the long term, remember some of these important aspects when deciding a dog to pick:

- Breed-specific illnesses
- Dogs with separation anxiety
- Dogs more prone to health issues

Of course, we want our dogs to be healthy. Because of years of breeding, there are certain health conditions that some dogs have gained throughout the years. While it is incredible what we have been able to do for dogs including the way that breeding helps them have certain characteristics, it is also essential to consider how breeding has led to more serious health issues.

Larger dogs, such as German Shepherds, might be more prone to hip dysplasia. Boxers are also at risk for elbow dysplasia, among certain cancers and heart conditions. The same can be said for Labrador Retriever. Both Labs and Golden Retrievers are also more prone to be overweight, which could put them at an even higher risk for health-related issues. These dogs require a lot of exercise, but it is also important that we don't put too much pressure or strain on their bodies to keep their joints intact and healthy.

In addition to some of these health conditions, larger dogs are also known for not living as long as smaller breeds. Part of this might be because of the way that they have been bred. They grow so rapidly that it ages them faster in the process. The larger the dog breed, the less likely it is to have a longer lifespan. This is sad in itself but remember the impact it could have on your family. Some large dog breeds, such as Great Danes, might only live to be five to eight years old (Soniak, 2013).

Smaller dogs are known for living longer since they age much slower. They might live up to be 15 years old depending on the breed. However, this doesn't mean they aren't still prone to disease and illness. Some smaller dogs, like Yorkshire Terriers or Miniature Poodles are more prone to have issues with their gastrointestinal systems. Part of this might actually be because of the stress they experience! Some smaller dogs are more likely to have ulcerative colitis, which is when the intestine or colon, becomes inflamed. It is more treatable, but when untreated or unmanaged, it can cause more serious health issues.

Certain breeds, like Pugs and Bull Terriers, have been so bred throughout history that they are very likely to experience certain health issues. They might have trouble breathing, could become dehydrated or exhausted faster, and will even snore at night in some households (Kiefer, 2017)!

All these dogs are equally cute, special, and important. It is still crucial that we consider their medical needs, because vet bills can be massively expensive. If you can't afford to pay for certain surgeries upfront, you could stand to lose your pet. Some veterinarians will work with you on creating payment plans, but many will require that you have several thousand needed for surgery upfront. We should do everything to care for our pets, but if you are not in a position to pay $500 or more in vet bills unexpectedly, it is important to reconsider getting a pet.

Aside from issues with health, some pets will be a little needier than others and require more attention and care. Some have separation anxiety that could cause them to have deeper issues. No dogs enjoy being home alone, but they will get accustomed to it in due time and with proper training. However, sometimes, being left alone for hours at a time can cause them to have some serious anxiety which can mean they may have accidents indoors, chew at furniture or objects, and whine or scratch at the door for hours on end. Not only is this annoying for neighbors, but some dogs have actually scratched at doors to the point that their paws will bleed. Consider your lifestyle and ensure that you don't get a dog that doesn't do well snoozing on the couch alone for a few hours while you are at work. If you do get one of these dogs, make sure you have it in your budget to hire a dog walker or sitter to check in on the dog during the day while you are at work. Listed below are the neediest dog breeds that do not like to be alone:

- Cavalier King Charles Spaniel
- Australian Shepherd
- Labrador Retriever
- Jack Russell Terrier
- Border Collie
- Miniature Poodle
- Bichon Frise

Chapter 2: Preparing Your Home for Their Arrival

Your dog will no doubt be scared when you first bring them home. They're coming into a new environment they've never been in before, which can be rather frightening! Set them up for success by using care and consideration when setting up your home. Whether your dog is three weeks or three years when you introduce them to their new space, there are a few things to consider. Not only is the environment in which you'll be training them important, but the tools that you have are special to consider as well. Stay positive, create healthy habits, and hide anything you don't want in the mouth of your pup! Let's dive deeper into these topics so you can be as prepared as possible.

Starting Positive Habits from the Beginning

From the moment your dog walks into their new home for the first time, you should do your best to train them to have good habits. If you don't start right away, then the dog won't be able to pick up on these healthy habits you are trying to instill so you can elicit good behavior. Dogs are very ritualistic animals and prefer to have a routine. Not only is it their way of operating within our society, but it also gives them peace of mind. Dogs can be very anxious creatures, especially if they can't predict what will happen next. A routine gives them the chance to understand what they might need to be prepared for and the ways that they can best go along with the natural flow of things.

Don't put off any vet appointments and necessary shots. Your puppy might seem perfectly healthy with no issues whatsoever. You might not be an active person or someone who would take your dog out in public anyway. In this scenario, it can be easy to think, "Nothing bad is going to happen, the puppy will be fine." Chances are, this is true. However, after all the love you give to your dog, this should be an extra step to ensure that they will be around for a long time. You never know what other dog they might interact with who could spread something. They might get one flea bite which could have been from an infected insect. There's no point in risking your dog's health, so keep up with necessary vaccines from the moment you bring them home.

Make sure to purchase all that is necessary before getting started. The dog should be the last addition that you have for your home and family. They should have all the necessary toys, bowls, leashes, collars, harnesses, beds, and everything in between to ensure that they will feel comfortable in their new home.

Practice what it will be like to have a dog before you even bring your four-legged friend home. It can seem tedious, but it is a good way to ensure that you are really prepared for the routine. Start by making sure that you keep all your belongings picked up. Puppies are like toddlers and can be great at finding random objects to chew on or play with. Every three to four hours, make it a point to go outside for a quick walk. Make sure that you get in the habit of establishing set times and routines so you can leave the house for about 10-15 minutes every three to four hours, just as you would once you start walking your dog on a regular basis.

Check-in with your emotions, because dogs require a lot more than just physical energy. Make sure that you are a patient and compassionate person. If you are someone who struggles to keep their cool, especially in stressful situations, it will be good to come up with some coping mechanisms because puppies will certainly test your patience. There will be moments where they have accidents, they won't always fully understand what you are trying to say to them, and they'll cause frustration among some family members. This isn't a bad thing, but it can be challenging to deal with, especially for individuals who don't have their emotions in check.

Be prepared to put in the work all day long. Even though your puppy will have plenty of moments of playing, sleeping, and just relaxing, you will still have to be alert. If you can't see your puppy, you'll have to find them to ensure they didn't get lost under a piece of furniture or in a different room. You'll have to be prepared to wake up early to take them out, fill their water bowl, and feed them. If any of this sounds challenging, then there needs to be more preparation done on the owner's end before bringing a new puppy into the home.

Make sure that you have the proper discussions with roommates and other household members. Decide who will be in charge. For example, if you have a roommate who you aren't very close with and who isn't really interested in dogs in general, you can't expect them to be a caretaker. If you are bringing a dog home to a family with a husband, wife, and three kids, then you'll probably want to split the responsibility equally among everyone. It is important to have a talk about all responsibilities beforehand. You can create a schedule for family members. Maybe each day, a new person is responsible to take the dog outside. Perhaps one person is in charge of feeding the dog and everyone else has to take turns walking the dog. You have to have these discussions before, so you are prepared instead of being stressed and confused in the middle of it all.

There has to be consistency when training the dog. That is key. You'll have to make sure that there are house rules. Mom cannot let the dog on the couch if dad wants it to always sit on the floor. The kids cannot give the dog any snacks that they're eating if the rule is that the dog only eats dog food. One family member cannot have a rule that the rest of the family members don't follow because this will hurt the dog in the end.

Traveling isn't going to be easy anymore. Make sure that you don't have trips coming up, and if you do, ensure that there's someone there who can take care of the pup while you are gone. Make sure that you have a schedule so that the puppy is never alone for more than a maximum of three hours. For puppies younger than 12 weeks, it should really be no more than an hour. Once they are older than a year, and in some cases six months, they can stay home for longer and up to six to eight hours depending on the dog breed.

There will be a lot of changes. Nothing will ever be the same. Some things you'll be prepared with, others will certainly surprise you. Don't underestimate just how prepared you have to be!

The Right House Setting

Your home is your puppy's home. It has to be a welcome and friendly environment in which they can thrive in. Imagine that you are the puppy before bringing it into your house. You are currently with all your brothers and sisters somewhere trying to figure out how to walk, run, and even bark. Then, some new people you never met pick you up and carry you away. They bring you to a place that you've never been to before. What would you want in this new place to make sure that you feel right at home? We can't project emotions on puppies like humans, and animals don't stick with their families all their lives anyway. What's most important to remember is that they will be frightened because things are suddenly different from how they just were. Ensure that your puppy is given all it needs to feel right at home from the moment you bring them home.

The puppy needs its own space. As it grows, you won't have to dedicate an entire room to the dog. You might be able to eventually just have them roam the house, even when you are not there to watch over them. As a puppy, you should set aside a dedicated area where they can run freely, sleep, and nap during the day as you are with them. Don't give them free range of the house just yet. That will be too overwhelming, and they'll end up getting themselves into a place that you didn't even think of at first. Your puppy will want something familiar, so use baby gates to block off a section of the house that they can feel the safest in.

Give your puppy a place to play that doesn't have a rug that you don't want ruined. You can't bring a puppy home and into your freshly carpeted living room, only to get mad that it immediately pees on the carpet. If you do have a carpeted room, invest in some thrift store sheets or extra rugs to put over the floor to protect it from the puppy. At first, your puppy won't defecate or urinate in massive quantities that are hard to clean. It is still important to be prepared for them to go 'potty' wherever and whenever they have to as you are starting to potty train them.

You'll want something steady and constant for it to bond with. Create a little room in an office, a bathroom or laundry room for your pup. These are smaller areas that they can go in for naps and get away from it all. Block off stairs just like you would for a newborn baby. Anywhere that a small creature could get hurt is an area that needs to be entirely inaccessible to your puppy.

Use puppy pads to create a path to the back door or the door that they'll be regularly going out to go to the bathroom. Let's say you are creating a space for your puppy in the sunroom that is adjacent to your living room since there is a direct path toward the backdoor on the other end of the house from your living room. Additionally, you spend most of your time hanging out in the living room, which will make your puppy feel less lonely when they spend time in the sunroom. You put a little bed in there, their crate, and some water and toys. Use newspaper or purchased puppy pads to create a path all the way from the sunroom to the backdoor. Each time they have to go out, let them walk along this path so they get a feel for where they're supposed to be going to the bathroom.

Make sure other animals have their own spaces during this time as well. Your cats will especially be more scared, so you will want to put their litter box or food bowls in a private space so that they can have alone time as well. Don't do anything that will freak them out, but if their litter box is in the laundry room where you also plan to keep the puppy, then they might not want to go in there, and could end up peeing or pooping in a random spot in your home. Consider all the animals and their need for their own room, not just your puppy.

Tools Needed

Once your home and your family are prepared to welcome your new furry companion, it is time to start stocking up on the right supplies. The best place to purchase supplies are from a certified pet store. You can shop online as well. When you go to certain big stores that stock cheap and generic dog supplies, you might discover that the products aren't as high-quality. This is fine if that's in your budget, but there should certainly be an emphasis on purchasing the highest quality items for your dogs. It may seem expensive initially, but if you really think about it, you will only need to buy most of these supplies once. For example, you'll purchase food and water bowls for them, but it is not like you'll be buying a new bowl every month. Every year or so, it is good to replace things just because you never know what bacteria or other germs they're holding onto that don't get washed enough, no matter how much you clean and disinfect them. However, most of these things will be a one- or two-time deal to buy, so now is the time to splurge and spend your money on your four-legged friend!

The first thing you'll want is a collar, harness, and leash. At first, puppies won't be able to run far away from you, but this doesn't mean you shouldn't still train them on a leash early. Make sure that you buy a collar specific to their size. You should be able to fit two fingers sideways into the collar to make sure they have enough room to breathe and grow. Get them the appropriate tags as well. Their name, along with a phone number and even your name is a good choice so that others can call them and call you if the puppy ever gets separated and lost. Tags with their vaccinations will also be helpful so that others know if the puppy is at risk of carrying or catching any diseases or illnesses.

A harness is a good choice as well. Puppies especially will tug and pull on their leash, and only having a collar might mean that they hurt their throats frequently. As you start to go for walks, the puppy might frequently get into things they shouldn't. Maybe there's an old sandwich or chicken bone that was tossed on the street from someone who was walking by earlier. This is a big find for your puppy and they'll try to devour the snack! Rather than jerking their necks back on a collar to stop them, you'll have a harness that could even help you to pick them up quickly.

Training pads will be a good addition as well. These are items that don't necessarily need to be the most expensive brand, and many stores carry generic versions that do just as well. The point is that they absorb moisture and provide your puppy with a scent to let them know that this is the place they should be using the bathroom, nowhere else. These are not to be used in replacement of going outside. If you want to train your puppy to use pads even into adulthood as they become full-grown dogs, that's perfectly fine. Some people might live on the 20th floor of a building and going outside three to four times a day isn't an option, especially late at night in the big city. They might want to use pads, and that's fine, it is totally up to the individual. However, if you do plan on training your dog to only go outside, the pads should be for accidents, not the only way that you teach them to go to the bathroom. They should be used during a transition period as the dog becomes accustomed to going outside.

You need to also purchase dog food and treats, but we'll get into the specifics of this later. As far as food bowls go, stainless steel is always the best option. Plastic bowls can trap moisture and cause bacteria to grow, potentially making your dog sick or even giving them little pimples on their chin! If you are using a glass or porcelain bowl, it could end up breaking if your puppy gets too rough. Stainless steel is best, especially since it can be cleaned very easily. Purchase one that has a rubber bottom to keep it from sliding all over the place as your dog eats. Keep it low at first for the puppy, but depending on the breed, a raised food bowl might be better so that they don't have to strain their neck or back to reach down to eat.

Have some form of a pet bed for them. It doesn't have to be a big dog bed, it could simply be a cardboard box with some blankets and a pillow. You could give them a few different options as well to see which bed they like the most. They simply need a place where they can sleep all alone and do so whenever they feel like it.

Toys are a big requirement! You want to teach them to play with certain objects and not to play with things that are more important to you. We'll help you discover the way that you can find what's best for your furry friend. Let's discuss now the option of using crates as a training room for your dog.

The Training Room for Your Dog

You have two options for your dog's 'room.' You can use a crate or you could have an entire room dedicated to them. A crate is preferable to most families, but if you can dedicate an entire room such as a spare bedroom, this is always ideal. Don't use something like a mudroom where people will be passing through frequently. For this section, we are mostly going to be discussing a crate as your dog's training room.

The reason that your dog needs a room is because they are den animals. This means that they are used to finding small holes and caves that they can associate with being their own. Think of animals in the wild that are like dogs. Wolves frequently find caves which they can easily hide in. They might find an old animal nest and use this as well! Your dog's room can be like their crate. If you choose to use a crate, then you will want something much bigger than the dog. For example, look at the photo above. It looks like it could probably fit into an even smaller crate, but it shouldn't. The dog should be able to stand up and fully turn around in their crate. If they can't, it is beneficial to purchase something a little bigger for them.

This room is not something that you should use to simply shut them away in. It has a specific purpose. This is the room your puppy will use to relax, play, eat, and sleep in, along with their puppy training. It is not a form of punishment and it should never be something that you use to isolate them. It is a safe haven, a place they want to go, and somewhere they feel completely comfortable in.

Book 3 - Training Your Puppy Step-By-Step

This is the room that they will go to when they need to sleep or when you aren't going to be home. A place of comfort and familiarity. If they have been playing for over an hour and you trained them throughout, you can put them in the crate as a way to let them know that this is a quiet time.

Never use their crate as punishment. If they go inside the house, chew on something they're not supposed to or engage in any other negative behavior you want to punish, don't put them in their crate. They won't be able to understand that what they did is wrong so therefore they need time to reflect and think about things. They will simply see that you are upset and confused about the situation.

Don't put your dog away for more than three hours at a time when they are younger than six months old. As your dog becomes an adult with more bladder control, then you can crate for up to eight hours, but this should really be the maximum. This should only be in rare circumstances when you can't find someone to help them out, not the entire day. If you are someone will be gone for eight hours and beyond, using a crate shouldn't be your puppy's room. You can use the crate and give them the option to go in, but a bigger space is better if you ever have to leave them for that long at a time.

Your dog wouldn't go to the bathroom in their den, so don't expect that they'll just go if they really have to. Instead, they will hold their bowel movement to the point that they physically can't anymore. The den isn't a place to put them to go to the bathroom. It is an area where you can train them to sleep, eat, and relax. If you want to leave them home all day and don't care if they roam freely in your home, then using an entire room with puppy pads is a better option to keep them away.

This is not prison. Only keep your dog in its crate as you would want to go in your room. If you make them fearful of the crate even once, they will be scared of the crate every time that they go in it afterward. This is a place where they can be alone, have their private time, and stay completely relaxed.

Even if you want to sleep with your puppy, they should go through a crate training period. This is to help them control their bladder. At night, you can put them in the crate and be sure that they will try to hold it in all night. If you let them sleep with you, they might jump off the bed and wander to a spot to relieve themselves, then come back to bed without you ever knowing (until you find the mess, of course)! After a few weeks of crate training at night, you can start letting them out of the crate and see if they are able to sleep with you while holding in their bladder movements all night long.

At first, this room should be somewhere they can easily access where you and the rest of your family will be. Don't put their crate in the basement where they can't get to it. Keep their crate in the living room where everyone hangs out or the kitchen where you might always be cooking. This is because you don't want to create the idea that their crate means isolation or that they can't be around others. If you get them used to them being in their crate while you are still there, it will help them better understand that this is their home, not their punishment. Now that you understand the importance and meaning of this small space, let's discuss how to introduce them to the crate. Again, if you don't want to use a crate, a small room is just as fine. We'll be discussing how to use the crate and introduce this idea to your puppy, but by all means, apply this to a smaller room as well.

You'll start by introducing this crate to them just as you do everything else – slowly and led by them. Put some toys in there and let them go in on their own. Maybe you place a towel in the crate that has the texture of their mother to make them feel more comfortable. Some people have even tried to put clocks in there to mimic a heartbeat. You might even put a treat or two in the crate. Let them go in on their own and explore this place for themselves. It will be a place they can get to know and explore, not just an area where they are shut away.

Never force your dog in the crate if they are actively resisting. If it seems like it is something scary, be patient. Sit next to the crate and play with toys. If they aren't into the crate, don't pick them up and put them in. If they go in just with one paw at a time, don't push their butts trying to get them to go faster. It might take a few days for them to actually go in, but that's totally normal. It will be better for you and your puppy if you are patient in this process. If they don't go in, place a toy at the front of the gate, not all the way at the back. Let them get this toy out if they want, and even if they don't go in the crate at all, it is still helpful to make sure that they are slowly transitioning into this.

You can make this transition easier by giving them meals in their crate. Start by introducing them to the crate on the first day. On the second or third day, once they display more curiosity, put their food bowl right inside the crate, just at the front of the door so they could even eat their food without going all the way in. Each time you give them a meal, push it a little further from the door so they have to go deeper into the crate. Eventually, you can put the food all the way to the back of the crate.

This is when you can start to close the door while they are in the crate. Don't slam the door and startle them. Slowly close the door behind them and see how they react. If they panic, open the door immediately and use positive reinforcement once they come out. If they don't even notice, congratulations!

Keep your dog in for 10 minute intervals, but nothing longer than that. After they've finished eating, let them stay in there for a few minutes, then let them out. Do this during every meal time until they don't show any signs of being afraid of their crate. Start leading them in without meal times after this. For the first week, do so every day. Practice letting your dog in the crate and keeping them in there for at least five minutes, but not much more than 10 minutes. This is to get them used to it. You should always be there in the same room during this training period. You don't have to be engaging with your dog. In fact, you should do the opposite. Watch TV, read something or just relax on the couch by them. Let them see you, but don't feel the need to look at them and play with them while they're in the crate. After these few minutes, let them out and use encouraging reinforcement the whole time, so they know that this crate is a positive experience.

Don't let your dog out because they are whining. Wait until a few minutes have passed after they've stopped whining to let them out. If you don't wait, they will associate whining as their key to getting out of the crate! It is hard to watch your puppy cry, but it is okay! It is better for them in the long run.

Practice this crating method before you actually leave them for long periods. Until they're used to a crate, you'll want to be home with them. This is why you should take a few days off work to make sure you are home with the new puppy.

You should be crating them when you leave and when it is nighttime, but not only these times. Crate them when you need to clean the house, or if you need to get some work done. Shorter periods are better, but if you only put them in the crate when they're alone, then it will make them associate the crate with isolation (The Humane Society, n.d.).

Hiding Certain Belongings

Puppies are like babies. They will explore and try to put things in their mouth as often as possible. It doesn't matter if it is a $300 pair of shoes or some poop from the litter box. If it has a scent, the puppy will try to see if this is food! You have to watch your puppy like you would babysit a child roaming around your house. A puppy has heightened senses, so they'll be even better at picking things out that shouldn't be eaten!

Make sure all electronics are picked up. This includes wires connected to the TV. Any cords or cables freely around the house should be put up. Use plastic hooks that can easily be removed from your walls to help keep wires and cords off the ground. Not only is this so they don't chew on them, but you never know what area of the floor might have a scent that triggers them to relieve themselves right there. If you can't put the cords up for whatever reason, you can use tape to keep them stuck to the floor and protected from your pup.

Block off electrical outlets. If you can't use a specific plastic child/pet safe outlet cover to protect them, use a baby gate to create a barrier to heavy electronic areas in your home.

Ensure that anything you don't want in your dog's mouth is put away. This is anything from shoes to throw pillows and chair legs, and so on. If you don't mind a little puppy slobber on your furniture, then keep it around, but anything else, make sure to hide from the pup. This won't be the case the whole time. Once your dog learns some rules and boundaries it'll be easier to ensure that they aren't getting making a mess or chewing on what they shouldn't be.

Check for loose buttons on furniture, thread getting pulled out, and other things that might be easy for your dog to put in its mouth.

Consider the cleaning products you might have used on the surface of certain things. Did you just bleach the bathroom? Did you use wood polish on the furniture? Your puppy might try to chew on these surfaces, and you don't want them accidentally licking up any harmful chemicals.

Plants can be an area of interest for pets. Ensure that any plants within the reach of your puppy have been put away, or that they are at least safe for animals to chew on and not something specifically poisonous to dogs.

Keep other toys that are meant for bigger dogs away from your puppies. This might be bones, big rubber items, and anything else that can be hard on their teeth or cause them to play too rough. Puppies need puppy toys! Consider things that you would never assume your dog has interest in that they still might. You might think, "That's fine sitting there, why would my puppy have any interest in that?" only to discover that it is in their mouth five minutes later.

Of course, there are many toxic foods that your dog should stay away from. These include:

- Chocolate
- Caffeine
- Grapes
- Onions and garlic
- Alcohol
- Raw meat
- Meat bones

Dogs love to chew on bones, but these should be safe bones purchased specifically for dogs. Chicken bones are more brittle, especially once cooked, so they could snap in half and your dog might end up choking on it. Even if you think you have managed to pick everything up you need to, you might discover that the puppy helps you see things that weren't there at first.

Choosing the Right Toys

All dogs deserve to have plenty of fun toys to play with! Toys help them learn how to interact with others while also providing them some entertainment. Dogs need to feel like they have a purpose. They like tasks because it makes them valuable members of their pack. They prefer to be outside chasing mice, birds, and other little critters to bring to their pack, but that's not ideal for us as human owners! Provide your puppy with all the toys they need for the best development possible. The best option for purchasing toys is to do so at a pet store. Be mindful of the quality of toys you purchase such as durability and materials.. Cheaper toys might have plastic or fragile rubber that breaks apart after just a few minutes of chewing.

Book 3 - Training Your Puppy Step-By-Step

Don't assume kids toys are safe for dogs. Just because a baby can play with it doesn't mean that it will be safe for your dog too!

Use toys that are specifically toys. Don't give them an old pair of shoes just because you don't care about the shoes anymore. They will think that all shoes are things that they can play with, not just the specific pair that you gave them.

Consider toys based on what your dog's breed is. Some dogs are hunting dogs that care about catching game. Toys that mimic animals like birds and ducks could be good for them. Other dogs are more interested in running and chasing, so they might do better with balls and toys that you throw long distances. You can never go wrong with a game of fetch! Many dogs like the same kinds of things, but it is also important to check with your specific breed and understand what their favorite things to do are.

Puppies want soft things they can roll around with. Their teeth are still developing, and they'll also want things that mimic their brothers and sisters. Puppies play with their siblings to learn how to function as adult dogs. They need that interaction to help them learn and grow.

Before they're six months, they will be teething pretty hard. Rubber toys and other things they can chew on will be best for them. Older dogs should have softer toys as well, as their teeth will be more sensitive.

For large dogs, don't give them anything may break right away. Even though it might be a specific plush toy made for dogs, your pup might have tough enough teeth to rip through anything they want.

Give them a variety of toys, not just one. This will help you discover what they like and dislike as well.

Dogs can have puzzle games too! You might get them a ball that has treats in it, forcing them to move the ball around to have treats fall out.

It is important to have toys that they can play with, however you want to ensure they socialize with kids as well and can play with them too. Some ways to ensure this are by including ropes that you could play tug-o-war with and having your kids throw some balls that they can fetch. Now that you have all the supplies you'll need to help your dog grow and develop, let's look at the actual training practices you will be teaching your dog.

Chapter 3: Principles to Train Dogs

You have your puppy, you have all purchased all your supplies to ensure your puppy is happy and comfortable, and have gone through any other measures to ensure they are healthy and safe. You are prepared. Let the training begin! There are a few essential principles that will be important to teach your dog. You should follow these rules and do these things as best as possible to give your dog a chance to be as happy and healthy as can be. Potty training will be the first thing you teach them. This is because it is a basic need that they have, and there's no delaying their natural urges. In this chapter, aside from potty training, we will give you a few basic training rules as well as methods of positive reinforcement to use throughout the training process.

Potty Training

This is the most essential thing that your dog will learn. We will spend a lot of time in this section, because this is one of the biggest headaches for new dog parents. A dog that chews on some furniture isn't a big deal, especially because puppies haven't developed strong teeth yet. Having to teach your puppy some boundaries can be simple as well. The biggest headache is when a dog goes to the bathroom inside the house. It is stinky, inconvenient, and generally annoying. Don't worry or fear this though! It is still natural, and plenty of dogs are easily trained to the point that they don't urinate or defecate inside at all and haven't for years at a time.

There are a few important things NOT to do to make this transition much easier you are your pup. First and foremost, never push their face in pee. This idea had been around for a long time as a way to train your dog. People thought that if you made your dog uncomfortable in this way they wouldn't go to the bathroom inside the house again. This is completely false and could even delay proper training. You are simply teaching your dog that you are dominant and that they should be ashamed for going to the bathroom.

Making sure that your dog is properly trained isn't just about taking them out quickly when they have to go. To help prepare them and set them up for success with potty training, ensure that they are on a consistent schedule and your communication is consistent as well. Dogs need regulation. They don't know the difference between inside and outside. They know the difference between going to the bathroom in an area that's acceptable and somewhere they shouldn't go. The only way that they'll know what's right and wrong is by you teaching them the boundaries they have to follow.

You'll want to ensure that you help them have a schedule by feeding them at the same time every day. If you are feeding them at the same time every day, this also means they'll likely be going to the bathroom at the same time every day. It will be easier to predict their schedule so that you know when they need to be taken out. The dog will appreciate this as well because it will make it easier for them to hold their bladders when they have to.

Having a crate is the best option in the process. They don't want to soil in their cage. They like to keep their environment clean just as humans do. Cleanliness isn't about appearance. Most animals are aware that it is simply not good for your health to sit in urine or feces. Never let them go in the crate or expect that they will go in their crate just because you leave them in there for a long time.

Your puppy will be able to hold it for as many hours as they are in months, until they reach nine months old. This means that:

- A two-month old puppy can hold it for two hours.
- A three-month-old puppy can hold it for three hours.
- A four-month-old puppy can hold it for four hours.
- A five-month-old puppy can hold it for five hours.
- A six-month-old puppy can hold it for six hours.
- A seven-month-old puppy can hold it for seven hours.
- An eight-month-old puppy can hold it for eight hours.

Eight hours is the maximum that you would want your puppy to be holding it for. Anything longer could hurt them. They might be able to hold it for longer, but that doesn't mean that it is healthy for their body overall. Some puppies do take longer, so don't be upset if your six-month old puppy can only hold it for four hours or less.

You will want to feed them in smaller bits throughout the day rather than one big meal at once. Puppies should be eating around four times a day, and as they grow, you can cut it down until they are eating two meals a day. We will discuss food and feeding in a later chapter. It is simply important that you are making a list of everything your dog does. Even if it is something as simple as:

- 8 AM: peed outside
- 8:30 AM: ate
- 10 AM: peed inside
- 10:20 AM: peed and pooped outside
- 11 AM: ate
- 12:50 PM: peed outside
- 2 PM: pooped outside
- 3 PM: ate
- 4 PM: peed and pooped outside
- 6 PM: peed inside
- 8 PM: ate
- 10 PM: peed and pooped outside

As you create this schedule, you will notice a pattern. Then, it will be easier for you to know when it is best to crate and when you shouldn't, for example, you might crate them around 9:50 or 10 PM for a little bit as it seemed like they went to the bathroom within a shorter time interval. This is an instance where they need to be bladder trained, as they probably could have held the pee that they released inside at 10 PMuntil 10:20 when you took them out.

The best method of potty training is to simply take them outside around every two hours. Take them outside before feeding because they might not want to eat if they have to use the bathroom really badly. When they do go outside, use positive reinforcement. Take them outside as often as you can, so if you can do so every hour, that is even better. Give them a treat, and verbal praise along with a pat on their head or rub their belly so they associate going to the bathroom outside with good things.

If your dog does soil the house, do not yell at them. First, try to distract them. Maybe you call their name, clap loudly or squeak a toy. Distract them as you notice them about to go, and then take them outside as quickly as possible. If they do end up soiling your home and you discover it moments later, simply clean it up as best as you can and take them outside. If you yell at them as they are going to the bathroom, even if it is indoors, they will associate going to the bathroom as being bad, not just going to the bathroom inside as something that's bad for them.

When you see the behavior, try to make sure that you stop it. That is the only thing you can really do. This is why some people get frustrated. They want a way to punish the dog immediately in order to prevent it from happening again. The only thing you can really do is make sure you are taking them outside as often as possible and stopping them with distractions when they do want to go inside.

Clean it as best as you can to get rid of the scent. If they urinate on a certain part of the carpet, you have to do everything in your power to get this scent out or else they might go right back to it to pee once again. If any other animals have ever urinated or defecated in the house, then you'll need to make sure that this is cleaned up as best as possible as well. Dogs have amazing noses that can pick up on a lot of various scents. If you aren't careful with clean up, they will be able to pick up the scent and think that this is the spot they're supposed to go to the bathroom. There are plenty of specific cleaners that you can use which are made for the exact reason of covering up another dog's scent.

Take them on regular walks. When they go outside, give them praise. This is when you could use a clicker for training. A clicker is a handheld device that makes a clicking noise. Give them a treat and then press the clicker. When they go outside and go to the bathroom, use the clicker and give them a treat. It is a way to keep them focused on the good things that they do. They will know that the clicker means they did something good. When they go outside and hear the clicker, this means that they deserve praise. This is a method that works for many, while for some, it doesn't work at all. Nonetheless, it is a way for you to try to better teach your dog all about positive association.

After you've been taking them outside for a while, examine their stool and urine as well. Is their urine in small amounts and bright yellow? They could be dehydrated. Is it a ton of water and extremely clear? They might be drinking too much water too fast rather than sips spread throughout the day. Is their stool runny and hard to pick up? Is it hard and giving them trouble as they try to poop? These are all going to be signs that they need different foods or treats. It can seem tedious, but it will be really helpful to have a strict schedule you use to keep track of when they're going and what their stool looks like (Bovsun, 2019).

Basic Training Rules

There are a few commandments of training that are important to remember. We will discuss the important things that you need to always keep in mind not just for training, but about dogs in general. They are a different animal, and that's what many people fail to realize. We understand that they can't do certain things like drive cars, walk on two legs, and get jobs. However, too many people put human emotions on dogs. We think that they have the ability to feel guilt, reflect, learn the boundaries that humans abide by, and so on. While they can learn rules, they won't be able to emotionally grow like humans. When you make a mistake, you think to yourself, "Ah, I shouldn't have done that." You might blame other people or you might take responsibility and change for the better. When you go through that situation again, you'll know what to do to not make the same mistake. A dog doesn't have this thought process. They're mostly thinking that one thing equals bad and another thing equals good. They will simply focus on doing things that are good, things that please you because you are the leader of the pack. All they want is to earn your respect. We can't read a dog's mind, so who knows, maybe they are better at existentially reflecting and growing. However, this won't help us in training them. We have to be patient and remember that they simply don't think the same way that humans do.

Remember that your dog does not have the same sense of time that we do.

Book 3 - Training Your Puppy Step-By-Step

They can't look at a clock and see how long it has been. Even humans don't always have the same sense of time in all situations. Think of how the ten minutes of extra sleep you get in the morning can go by in the blink of an eye but waiting in line at the DMV for 10 minutes can feel like three hours. Unless we actively monitor the clock, we don't always have a perfect sense of time. The same can be said about your dog. They don't even have the ability to look at a clock and measure time numerically. If you go to the store for an hour and a half, this can feel like the same time that it takes you to go for a ten-minute walk on your own. Dogs won't know the difference, so we have to remember that in training.

Dogs are not going to fight you back unless they feel as though their life is in danger. They can be naturally aggressive, but not because they are evil, vicious animals. They have a combination of instinct and feeling that will feed into their aggression levels. Dogs are only aggressive because they are scared. They need to learn how to play as well. If a dog is playing rough, they are simply learning the boundaries between play and real kill or attack. If they are aggressive and growing or snarling, they are likely just scared and trying to protect themselves.

If you are playing with your dog and they get to rough, make a loud and clear noise so they know that they are hurting you. Don't scream and yell, just a quick and sharp, "Stop" with an assertive over emotional tone is all that's needed. Some people will even yelp in the same way that a puppy would in an attempt to teach them about boundaries on their own level. The noise that you choose to make is up to you. Remember, you don't need to punish them. Simply stop the aggressive fight by letting them know they hurt you.

These dogs look up to you. You are their hero, their leader, the person in charge. Don't abuse this power that they've given you. Use it for good and train them to be happy and healthy dogs.

They will pay the most attention to your tone. Though they can't know exactly what you are saying, they will be able to pick up on your tone of voice to get a sense of whether or not you are happy or if you might be stressed or scared.

If you are stressed, your dog will be stressed. In fact, one study discovered that some dogs had the same cortisol levels that their human companions had. This means that they will pick up on the emotions we feel and then end up feeling it themselves.

Dogs are incredibly energetic, but they won't always be willing to run around and be animated. Remember that they need time to rest. Just because they don't want to get up to go to the bathroom doesn't mean that they're disobedient! We all know that feeling of laying down and not wanting to get up even if we do have to pee really badly. Whether your dog is a puppy or an old dog, they have the same kind of rules that apply to how you should be treating them.

Positive Reinforcement and Remaining Patient

You will want to become the pack leader for your dog, and any other pets that you might have. You can earn their trust through ways other than violence. Many individuals believe that physical punishment is the best way to teach dogs the rules. This is completely false, and studies have been proving that physical punishment over positive reinforcement actually hurts dogs in the end.

There is one important rule that you will have to live by among everything else:

<u>Never hit your dog</u>. This can't be reiterated enough. There are many people out there who might tell you otherwise. They might say, "It is their language," "It is how you become dominant," "It is how you show them who's boss," or "It is the only way they understand." Maybe it is a way to communicate. None of that matters, however. What we do know for certain is that dogs do not react well when they are hit. We know that it causes long-term damage, it physically hurts them, and they don't understand you. There are a million other ways to communicate with your dog. Screaming, hitting, kicking, and pushing your dog are not the ways that you are going to want to train them.

Know the difference between assertive and aggressive. When you hit your dog, it usually comes from an emotional place. Screaming might indicate frustration. There are emotions attached to these actions which dogs do not always understand. Separate your anger or frustration from the dog and do not take it out on them. Maybe they did poop and pee all over the house, got into the trash can, and chewed up your favorite pair of shoes. This is all terrible, but hitting them doesn't fix that, and it doesn't make them learn anything valuable. They will be scared as it is when they see that you are upset, and being aggressive toward only makes it worse.

Studies have proven that using punishment against dogs will increase anxiety and fear within them (Becker, 2012). They don't know that you are upset at them. They don't understand that you are trying to teach them a lesson. They will be confused and sad that the person they love more than anything is angry with them. Using aggression on them will make them aggressive in response to the same kind of stressful stimuli.

In fact, there is no increase in obedience in dogs who have been trained physical punishment as reinforcement for negative behavior Instead, some dogs are less obedient after experiencing repeated physical abuse. Use positivity for your training procedures. The way that they are going to learn better than any other dog is when you choose to use positive reinforcement with them. This means that rather than punishing them for the bad things they did, you'll reward them for the good things. If they are being bad, you can distract them and change their behavior by using a toy or another noise to pull their attention. If you don't catch a dog in the act of doing something disobedient, you have to wait until the next time to teach them this is wrong. If you get home from work and discover that your dog went through the garbage, they won't know that a punishment at 8:00 pm is for something they did at 6:00 pm. They have moved past this and forgotten.

Your dog should feel comfortable and trust you, not fear you. Put yourself in their paws, so to speak. Who do you feel comfortable around? Who do you love and trust the most? Who inspires you and encourages you? Who accepts you and loves you just the way you are, flaws and all? Then think of someone you are scared of, someone you cannot be yourself around, someone who makes you feel uneasy. Who is more effective in inspiring you? Which person do you appreciate more? Sure, your dog might listen to you if they are scared of you, but the bond you create with them will be built on fear, not love and trust. It is important to have a healthy and loving bond that brings the both of you together.

When training, use simple phrasing so that they can better understand what you are trying to say. Don't have complicated commands and explaining things in sentences doesn't help them. Simplify your talk. While we all love talking in long winded sentences to our dogs and they like to hear from us too, they don't really understand every single word except for relevant command words.

Ensure that you reward good behavior immediately after it happens. Don't wait too long to do this. Right after they've done something you approve of, show them immediately.

Be patient in this process. Don't overwork them. They have incredible power within their brains and can be very smart creatures. At the same time, they do have a mental limit on the day, and it is probably less than what we have. Don't expect to train them all day. Include breaks to ensure that the two of you focus on playing and relaxing together as well.

After they've started showing signs of good behavior on their own, you can reduce using treats to reward them. During the first few weeks, you'll probably need to use treats frequently to get them to sit, lay down, go outside, and so on. Once they are doing these things regularly, cut back on the treats that they have. Dogs older than eight months won't need a treat every time they go to the bathroom. Too many treats could mean that they gain weight and end up being at risk for other health conditions.

Act excited when they are behaving well instead of using treats every time. You could still give them treats every time if that's something you want.

Just ensure it is a healthier treat like frozen peas or carrot sticks.

Make sure to always finish up training sessions with a positive reinforcement. Even if you trained for 30 minutes and got nothing accomplished, end it by doing something that you know they're good at, like simply sitting. This ensures that they still feel good about training and that they'll be able to pick up where you left off next time you start again.

Studies have proven that the more engaged you are with your dog, the more obedient they are likely to be. If you take your dog on your runs and walks, let them ride in the car while you running some errands, and hang out with them while watching TV, they are more likely to be obedient rather than if you leave them at home or away in their crate during these times of activity.

Specific Rules for Small Dogs

Small dogs are great companions. They have small waste, they're always by your side, and they usually love to snuggle up right next to you. They are different from other dogs, so it is important to consider the ways that we should train them especially carefully to ensure their needs are being met.

What we have to remember is that the same rules apply for these smaller dogs, but we also have to take their size into account. They might have less energy to train for longer periods of time. They might be more passive and scared in larger settings because they are so tiny, especially as puppies.

Just because a Chihuahua might not hurt you when they bite you, that doesn't mean we shouldn't stop that behavior. We have to have the same rules for our small dogs as we do our big dogs. Even if their feces may be small enough to the point that we don't care if they go in the house, we still need to discipline them when they soil a place that isn't meant for them. However, if you have bad habits in one area, it can make more bad habits in other places.

Your smaller dog might initiate a fight with a larger dog, and that's when things will get dangerous. Though you don't care if your tiny Pomeranian is aggressive, it still teaches them that it is okay to be more aggressive. Then when they run into a Doberman on the street, they park and snarl, and the bigger dog fights back. This is where you would see the negative effects of improper training or condoning aggressive and negative behavior, so don't wait until things are heated to take action.

Train small dogs to do simple tricks when you are sitting down rather than standing above them. They can be easily intimidated by size, so make yourself as small as possible when working with them to teach them new tricks.

Don't assume that you should always be picking them up either. It is easy to just lift them up and plop them outside really quick rather than harnessing them up and walking them down the stairs. However, they need to learn the path to important places on their own, so walking them through rather than carrying is the best way for them to learn.

If you have a small dog and a big dog, make sure the small dog doesn't have more privileges than the bigger dog. You might let a Mini Poodle sleep in bed with you, but the German Shepherd has to sleep on the floor. This creates imbalances in power and the small dog might think they have more control, leading to unequal treatments.

Don't assume just because you are bigger that you are the one in charge! Small dogs can be known for having the most attitude, so remember that it is important to still maintain your role as the leader of the pack.

Specific Rules for Medium Dogs

Medium dogs are usually breeds that might be family friendly. These are the dogs that the whole family can play with, hangout with, and have a bond with. Medium dogs are the standard for what we discuss throughout the book, so they don't have as many specific training needs. Rather than considering their size, it is a breed that you will want to cater to.

Specific Rules for Large Dogs

Big dogs think that they are small dogs in some situations! They don't realize how big they are. Just because they jump on you and knock you to the ground, they aren't trying to attack you. Many people associate aggression with big dogs, but it is mostly just because they aren't aware of their physical impact. A small dog might get excited and jump up on your legs when they first see you, and you might hardly notice. A big dog will do the same, but their leap could push you back or hurt your legs, automatically making them seem more aggressive. Remember, big dogs aren't intentionally being physically stronger! It is just part of who they are.

Big dogs are going to be high strung, more hyper, and of course more physically boisterous when they haven't been given proper time to exercise and play around. Since they have a lot of energy they need to expend on a daily basis, they need a lot of time to play and run around, and they need plenty of space to do this. If you can't provide them with consistent exercise, a smaller dog is a better choice.

Just because they're big doesn't mean they can't still get scared! Even if you are five feet tall and you are training a 100-pound Great Dane, they can still be very scared of you. Treat them with the same love and compassion that you would a small dog.

Bigger dogs might need more boundaries than smaller dogs, especially so that they don't scare other people. Even though you might not mind it if your Greyhound jumps up on you every time you get home, visitors, especially children, could be scared by this reaction. Always consider your dog's size and breed when training and interacting with them!

Chapter 4: Training Your Puppy Outside of Your Home

You may be able to rein in your dog at home, but in public, it is a whole other story. Not only will your puppy be more distracted, but so might you! Training should never stop, which is why this chapter will provide you with tips to ensure your dog is doing well in public. This chapter is filled with all the tips needed to not only carry out training in public, but to specifically know the best methods of training in new environments. It is healthy for your dog to socialize, so even though it might be harder in the end, you have to participate in these methods to ensure your dog is growing in the best way possible.

Going for Walks and Leash Training

Even if your dog has a big fenced-in yard to run freely in, it is still important that you train them to use a leash properly. You will be keeping them on a leash as you take it to the vet, groomer, to doggy daycare, and so on. It is better to have them know how to use a leash rather than always tugging and pulling away from you. Your dog needs to wear a collar in case it ever runs away or gets lost so that it can find its way back to you. Having a dog that knows how to be obedient on a leash also means that it will be obedient in other important areas as well.

As soon as you bring the puppy home, get them used to wearing a collar. They don't necessarily have to wear it all day throughout the house, however, they should get used to wearing it at important times such as going outdoors. They might not like it at first, but eventually, they won't even notice it or feel it. It is important to get one that is the right size and strength for your dog. This is something that you will want to spend the most money on compared to other products within this category that might not be as important. You don't want to get a cheap collar that breaks easily!

A collar is necessary to help identify the dog, should it ever get away. There will always be a level of unpredictability when you have a dog, no matter how responsible you are as an owner. Your dog might be completely relaxed as you sort through your mailbox but then sees a squirrel for the first time and runs away as fast as possible, completely shocking you as it gets away! It can happen, so a collar and leash are important no matter the temperament of your dog.

Remember, your dog will pull and tug at the leash, trying to walk as fast as possible. Retractable leashes are banned in some areas, so your choice of leash type depends on where you are located. A leash that is at least five or six feet long is good for the average person and dog. Longer leashes are better for small dogs and taller people. The leash needs to be strong enough to keep the dog back. Train your puppy in a way that teaches them not to pull the leash as hard. If you have a very large and strong dog, a chain leash might even be something that you consider.

It is certainly beneficial to use a harness while training your dog as well. This way, you can pull the dog back without hurting them. There are a few different collars that you might use aside from just a harness too. The most common is a simple collar with a plastic buckle. This snaps together comfortably for your dog.

Another popular kind is the **martingale collar**. This is for dogs that have the same size neck as they do their head. It will be easier to adjust these so that it doesn't slip off.

There are also **half-choke collars**, which are looser collars that will tighten as the dog starts to pull. This is to keep them safe while teaching them not to pull as hard.

Prong collars have pieces of metal that stick out so when the dog pulls on the leash, the prongs poke them and make them uncomfortable. This is beneficial for the owner and training purposes, but it does hurt the dog. Some dogs will be fine and stop pulling when they feel the poke. Others won't even care and will continue to tug the leash as hard as they can. Consider your dog and their level of curiosity before you use a prong collar.

You can also get a collar that wraps around their face, known as headcollars. These keep their mouths closed as to encourage them to not bark as much. It is still attached to their neck so that you aren't tugging them by their face. It is best for dogs that bark frequently.

It is helpful to have a harness connected to the collar so that the dog knows they need to heel. While you don't want to choke them out, a little discomfort can be enough to teach them how to properly walk next to you. You might just think that dogs can easily start to walk as soon as you put the leash on them. There are a few important steps when introducing walks to your dog.

Stick to one side when walking. Whether you want them to walk on your right side or your left side is up to you. However, try to pick just one that they stick to in order to make it easier for them to be obedient when you start to go for walks.

Play with them first to make sure that they are nice and tired. If you go for a long walk first thing in the morning, they will most likely be active and excited to see everything that's around them. It is better to make sure that they've been playing awhile and are more tired when it comes time to take them for walks.

It is important to start indoors so they can get a feel for what's ahead. Once you are ready to start training your puppy, put a leash and collar on them. When they are sitting next to you with their collar on, this is a good time to reward them. Give them a treat for sitting right next to you and not moving. Start walking from one point to the next. When they take the lead, stop and wait until they are sitting right next to you. Feed them a treat again.

If they pull, do not pull back hard. Simply stop so that they know to stop, make them sit, and walk ahead a bit so that you are leading. Feed them a treat and use verbal positivity when they act this way. It is okay if they get a little ahead of you, but never to the point that they are tugging on their leash. They should learn how to walk right next to your side in order to make sure they are obedient when going for walks outside.

Be happy and engaging with your dog. Don't get overly excited as this might cause them to be too excited and then they'll want to run ahead of you. Practice walking indoors as frequently as you can. Use a clicker instead of treats as you transition to walking outdoors. If they are really struggling to not pull and don't want to walk next to you, then you will have to take treats outdoors. Remind them that you have a treat so that they'll be more interested in you rather than interested in what's ahead. If you are exciting, your dog will be focused on you rather than letting their attention carry them everywhere outside.

Book 3 - Training Your Puppy Step-By-Step

Doggy Daycare

Doggy daycare is a great way to start to make sure that your pet is well taken care of. You can find experts that love dogs to help keep your pets happy all day long while you are at work. Not only is it helpful for you, but it is better for your puppy. They'll get a new and exciting environment, interact with other dogs, and come home tired, so that you don't have to exhaust them right away.

First, do extensive research into your doggy daycare. Make sure to read real reviews from actual people from multiple sources as well. Do your research on what they practice, what they believe, and the basic steps they take their pets on throughout the day.

Consider other amenities that will benefit your dog the most. Do they offer grooming and bathing? Do they make sure to train dogs? Do they feed them? Do they have indoor and outdoor space to run? All of which are important things to consider for your pets. They should have a thriving environment that stimulates them and fulfills their basic needs.

Ask if you can tour the place first! Many will be fine with giving you an interview opportunity. They might not show you where they keep the dog out of respect for other dogs and their owners, however, it is beneficial to examine what the kennel situation looks like for your pets.

They'll likely have a training period with your dog. This might mean that they try things out for a week to see if it is a good fit or if your dog should go elsewhere. Socialize your dog at parks and other public places first to see what their strengths and weaknesses are. Do they get excited and run up to other dogs? Does this interaction make them anxious? Aggressive? Overly excited? Keep track of how they interact so that you can help the daycare workers as well.

Check in with what foods they'll have. Do they provide food or are you going to supply it? Will they be doing feeding at all?

Make them aware of all your commands. You want to share with them the tricks your dogs know or are at least trying to learn so that they can help your pets training further. You don't want to give them two different commands for the same action because that will just confuse them.

It is a good idea to find a daycare that also does training. This way, your dog will have consistency. You can take training classes with the dog and they can have the opportunity to learn from professionals, and then you can be assured all training procedures will be the same at home and at daycare.

Make drop-offs quick and don't get too excited when you see them for the first time. if you prolong the drop off, it will make them more anxious and scared. It will make each drop-off harder and harder. Make it short and sweet. When you pick them up, they will naturally be excited. Don't make it so that they are even more out of control with their emotions. It is tempting to be like, "HI! I missed you so much!" with a screeching, excited tone, giving them lots of hugs and kisses. How could any dog parents resist? However, this will make these moments harder and harder each time because the dog will only associate positivity with you, and not with the trainers or caretakers they're with all day.

Parks

Parks are like a dog heaven. Whether it is a regular park or one that's specific for dogs, they provide a great opportunity to get some exercise while healthily socializing as well. Find a park that is close to your home which you can take your dog to regularly. It is good to find one in walking distance so that you can get more exercise for your dog before they even get there.

Check-in with all the rules allowed at the park. Are dogs allowed? Are they allowed off their leashes? You don't want to take your dog somewhere that they aren't even allowed! Remember that they will be socializing with other dogs and other people. Are they prepared for this? Do they have the skills necessary to be a present animal that doesn't attack others? Will they get excited and jump on someone? Will they bark and growl at other dogs? While it is important to socialize your pet, you also have to remember that there needs to be a level of basic training already instilled in them so you can diffuse any potentially challenging scenarios.

Consider your dog's anxiety and excitement level. Are they going to be freaking out and be overly excited? Are they going to be more relaxed and instead stay close to you?

Walk around with them before going to the park to reduce their excitement. If you take them to the park without having any physical exercise in the day yet, then they will be more excited than anything else. Don't wait until you get to the park to work them out. They need time to calm down before playing or else they might end up being far too excited.

If it is a specific dog park, then it is usually fenced in so they can be off leash. Other parks might have fences that allow you to take them off their leash as well but check in with the rules. You might also want to take them off their leash no matter what, but you have to ensure they are properly trained before ever doing this. If you ever take them off their leash with no fence, they have to know how to come on command, how to walk next to you rather than ahead of you, and their excitement with other people should be low. Even if this is the case, it is still not recommended to let them run free where there's no fence. It is not just for your sake, but for other people as well.

Don't take their leash off right away when you get to the fenced in park either. Give them a chance to sniff around. Let other dogs come up to them if they want and give them a chance to look around and get used to their surroundings. If you let them off their leash right away, they might start darting around and become concerned and fixated on the wrong things. Warm them up to this new public setting by keeping them on their leash for the first five minutes, minimum.

Don't build the excitement too much either. It is easy to say things like, "Do you want to go to the park?" in an excited tone. How could we resist seeing and showing off how cute our pups are when they get all excited. However, this is going to make them more active when you get to the park. They could be excited to the point that they're uncontrollable, and that's not what you want for your dog.

Book 3 - Training Your Puppy Step-By-Step

If you are taking a puppy that's not yet spayed or neutered, consider how other dogs may react. If you take your tiny Mini Poodle girl to the park and she's not fixed, she might get approached by a lot of larger male dogs. This could cause her anxiety, and she might even fight some back. Don't let this happen and keep her close if that's the case. If you have an unfixed male, he might run up to other females and try to bother them as well. It is best to have all adults fixed, but for puppies that are too young, keep them close.

Feed them long before so that they can go to the bathroom before you enter. If you feed them right before you go, they might poop right away when you get there. That's fine for dog parks, but for regular parks, this could be against the rules.

Pay attention to your dog and interact with them constantly. This shouldn't be the time to scroll on your phone. You never know what they might get into, and you will know even less what other dogs could end up doing. We can't assume that all dogs at the park have been properly trained in the first place.

It might be a good idea to leave the toys at home. A ball is usually fine, but specific toys could cause conflict. Another dog might approach your puppy, and they could get defensive over the toy. Dogs tend to be more competitive at parks. Remove the chance of conflict by leaving potentially triggering items at home.

Pay attention to what they might be picking up and sniffing at. Especially puppies. They might stumble upon a fresh pile of feces left by another dog and decided to give it a lick to see if it is a treat. The point is to have fun, so don't make it a stressful time for your dog! Be as prepared as possible to meet all their needs.

Friends' Homes and New Environments

There are a lot of people in the world who don't care about how other dogs act. They'll be excited to see a puppy and the first thing they'll want to do is shower it with attention, even if your dog might be misbehaving. They won't always be as serious about training because they aren't thinking of the dog's needs in the long run, they simply want to hold and cuddle a cute little puppy.

Though you might not care if your dog is jumping on strangers, remember that going over to someone's house might mean that they'd prefer if you didn't allow your dog to do that. You might have no boundaries for your furniture and the dogs are allowed to sit and chill whenever they want. Other people might not want your dog on their couch at all, so respect the rules of other people's homes.

You might have your friend be pet sitting as well. Whatever the situation, there are a few rules to remember. Keep them leashed when you first arrive. Just like the dog park, they are going to be overly excited, so you need to give them some time to warm up to their new surroundings. If you don't keep them leashed, they might end up running rampant around the new home to sniff out everything and anything!

Book 3 - Training Your Puppy Step-By-Step

Have a toy that will keep them distracted. If you are going to a friend's house to watch a movie or play a game, you'll want your dog to be focused on something other than you. Bring a bone for them to play with or another treat that will keep them busy on their own. This will help make sure that they are relaxed and free from getting into typical doggy trouble.

If they're staying the night, ensure that you have a place for them to sleep. Bring their crate with you if they are still crate training. You want to ensure the same consistency in night time routines at someone else's house that you have for your own home.

Check-in with any other pets in the house and what it might mean for all the animals. Does your friend have cats at his or her house? Has their cat interacted with dogs before? t They might need to keep the cats in a bedroom during the visit to protect both the dog and the cat.

If you are introducing it to a new dog, have your friend bring them outside. If you bring your puppy into a dog's home, this could make them defensive. Instead, have them meet on common ground so that they're more respectful of each other, and then you can bring them together.

Bring supplies in case they decide to use the bathroom in the house. While you might think you have it under control because your puppy hasn't gone inside for the past few weeks, this new environment could be the one place they decide to relieve themselves.

If you can, walk your dog through the house to the backyard or whatever the pathway might be to use the bathroom. Just like you would for your training at home, you want to show them the way to the bathroom so that they'll know exactly where to go when they need to relieve themselves.

Remember to be respectful of house rules as well. Even though you let your dog on the couch, your friend might not want that. This is why it is important to teach them 'down' commands when they're at home on the couch and bed. If they know how to get down, then you can use this command if they do end up jumping up.

What to Do When You are Not Home

You should take three or four days off when you first get a puppy. Unfortunately, there's no "new puppy" leave option for work. Imagine if we had puppy-leave like maternity-leave! After getting your dog introduced to the new home and when you have to head back to work, you'll need to use different methods to care for them when you are not home.

The first thing you should do is fill in help and reach out from friends and family to be with the dog. After six months, your dog will likely be in some healthy routines, though they're still technically a puppy. Before this time period, see if you have anyone, or a number of people, that are willing to come over and hangout or play with the puppy for a bit, or if they can at least stop in to take them outside fast. Many people will be more than willing to help out for puppies!

You can hire a dog sitter or take them to doggy daycare. There are many apps and other services that make it easy to find a dog walker in your area. They can send you pictures of the dog so that you can make sure they are being properly taken care of.

If none of these are an option, then you'll have to make sure to set the home up for them when you leave. Remember that they should never be in a crate for longer than eight hours, but this is the absolute maximum. That doesn't mean that eight hours and 26 minutes is okay. If anything, you should keep in mind that seven hours is your maximum threshold and you can go over that by a few minutes here and there, or on the days you are running late.

Take them outside before you go, but don't make it something that they are scared of. Treat it like normal. Leave your purse or other things you are taking with you inside so that they don't suspect anything. Just like with the doggy daycare, don't make them overly excited or else these will become focal points for your dog's excitement.

Wait around a few minutes after they go to the bathroom. If you take them inside immediately after, they will learn that peeing or pooping means it is time for you to leave. They might end up holding it for a while because they don't want you to leave. Then you might think that they don't have to go, so you take them back inside, only for them to have to hold it for even longer. They should have plenty of time to go to the bathroom.

Crating at first is fine, but only for shorter periods of time. Remember the month-to-hour ratio for how long they can hold their bladder. If you are going to be gone for a long time, you might consider giving them a room. This way they have several different options to lie around and play, and if they do have to go to the bathroom inside, they aren't doing it in their crate where they have to lay in it.

Tire them out if this is an option. If you can play with them before you leave for work in the morning, this makes it so that they can just focus on sleeping while you are gone. Invest in a doggy webcam so that you can watch what they're doing. Some even give you the option to press a button to talk to them or release a treat!

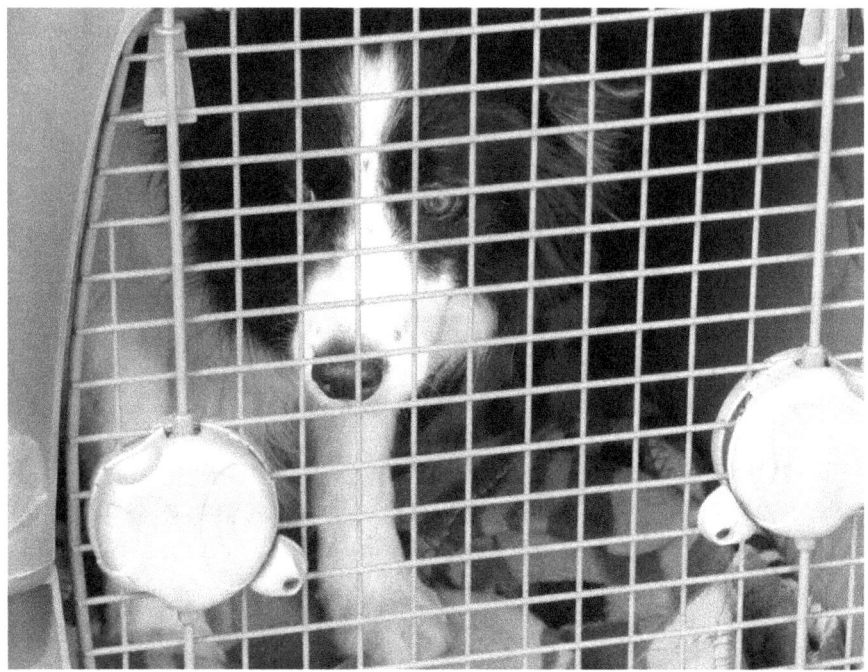

When you get home, the same thing goes for the doggy daycare. Don't make it too exciting or else this is how they react the first time they see anyone. Make it quick and casual and take them outside immediately. Remember that they have been laying around all day, so they will be rather energetic. After a long day of work, you may not be nearly as excited, so don't punish them because they're extra happy. Make it a point to play with them for at least 15 minutes when you get home from work, or wherever else you were. This ensures that they'll get tired out and you won't have to deal with a crazy energetic dog for the rest of the night.

Public Tips and Reminders

You might be getting a dog as a service animal. You might be someone who is more active than others. Whatever your reasoning, you will probably be taking your dog out in public frequently. We have a few more tips for you to remember when taking your dog out into the social world.

Book 3 - Training Your Puppy Step-By-Step

Keep your dog training even when in public. Consistency is key so you don't want to create an environment where your dog is free to do whatever they want just because you are no longer home. It will confuse them and make your training regress.

Just like the other activities we discussed, try to tire them out before going out in public. Establish some own boundaries with what you might say to those who want to pet your dog. You might not want others to distract them or make them excited, so it is perfectly fine to say, "I'm sorry, but we're in training right now," if someone asks if they can pet your dog.

Keep a few treats in your pocket so that you are not out of options should you have to reward good behavior. A clicker is a good option in public because it helps bring their focus back to where it should be without it being a distraction.

If they are acting badly, don't just give them a treat to calm them down. This will teach them that bad behavior will get rewarded. If you have one dog in training and one dog that's older and already well-behaved, try to refrain from having both of them out in public at the same time. This might make it harder to focus on training the one. Whatever you do, maintain consistency, and stick to the same rules no matter where you are.

Chapter 5: **Skills that Dogs Need to Know**

Though it is cute to teach your dog to sit and shake your hand, roll over or play dead, they also need to know some basic skills such as sitting and laying down. These are important things to teach your pup because it will be the way that you control them when needed as you are training. When you can teach your dog new tricks, they will be more obedient. Remember that these training processes aren't just to be done in the beginning, but all throughout your dog's life so that they're always effectively learning in the best way possible.

Sitting and Laying Down

You should start to teach your dog to sit as soon as they come home. Just because we're discussing this in chapter 5 doesn't mean you shouldn't start right away! Sitting is going to be one of the most important tricks that you will teach them. Even the most untrained dogs will still know how to properly sit. It is a natural state for them to fall into, which helps you have better control over them.

Remember positive reinforcement is key! Don't scrutinize them because they're not sitting. They won't understand why you are mad. Have some treats ready to hand out as you teach them this valuable skill.

Sitting is a great command because it is the way that you ensure your dog will be relaxed and chill when you need them to be. Maybe you are about to feed them and they keep running in circles around you. Perhaps new visitors are coming and you need them to be relaxed and sit down so they don't jump on others. You can command them to sit to ensure that they'll be as relaxed as possible in these times of great excitement.

Book 3 - Training Your Puppy Step-By-Step

Do these training sessions in shorter time periods. No more than 10 minutes is good. If you go for longer, it can tire them out. It can be too repetitive, and they might be desensitized, especially if they're not properly learning the different techniques.

To start, hold onto some treats. When they sit, give them a treat! It is as simple as that. You should use the normal command, 'sit,' rather than something more complicated. They might not sit on their own for a while, and that's fine. Simply give them a treat only once their furry bottoms are touching the floor and they're giving you attention.

After a few times, start to say 'sit' as they sit down. This is the way that you can associate the word with the action. Continue to tell them to 'sit' on a regular basis whenever you want them to be more composed. Each time they sit, say "good job" and give them a treat. You will discover that they know that if they sit something good will happen, so they'll do it more often. Ensure that you are also using positive reinforcement and that you have an overly happy tone.

Stick to this word because it is how others will know how to command them when you are not around.

Don't expect them to stand up again after sitting either. Step backwards or away to get them back on all fours, and then ask the command again. This is the way that you'll be able to teach them what 'sit' means rather than using your hands to force them to sit or to stand up.

Practice without the treats to make sure that they're getting it. Slowly phase out the treats after a couple weeks of this practice. They have to learn that not every time they sit they get a treat, but still use positive words so they know how good of a dog they're being.

They should learn to sit when:

- You are giving them something, like their food.
- Before you put their leash on to go outside.
- When someone arrives at your home.
- When they are being nosy and begging for your food.
- Before jumping onto beds/couches.
- Any other time you need their attention and focus.

Laying down is just as easy, but they need to know how to sit first. After you've successfully taught them to sit, you will be able to teach them so many more tricks even easier. It is like learning an instrument. Once you know how to play one, the rest will come simpler. To teach them to lay down, you'll have to start by paying attention to when they are actually laying down.

You'll do it in the same way that you taught them to sit. Wait until you see them plop down and laying there and give them a treat. Once they're laying down, give them praise. They'll likely stand up from excitement, so it is a good opportunity to practice over and over again.

As you see them start to lay down, tell them, 'down.' They will then know that this word will mean that they have to lay down on the floor. Another method to teach them to lay down is to put a treat on the ground with your hand over it. they will try to get as close to it as possible, so in the process, they end up laying down. Give them the treat a few moments after this so that they understand that it is the laying down which is giving them the treat, not that this is just a random occurrence.

Never force their back down. They won't understand what you are trying to do, and it won't be the way to teach them to sit. If you are just forcing their back down, they aren't learning how to do it on their own, so they will only do this when you are the one pressing on their back.

Bark Control

Barking can be a huge issue for dogs. This is one top reason that people end up taking their dogs back to shelters or give them away as well. To make sure that your dog's barking is under control, not just for your sake but for your neighbors as well, there are a few things you can do.

Dogs that herd and have other sports jobs might be more inclined to bark. They think that barking is the way of communicating and that it is part of their job. If barking is something that's going to be a big issue for you, it is best to first not pick a dog that is known for barking.

Look at the reason that they are barking so often. Is there a distraction that they always have to bark at? Maybe it is the mailman or other traffic that frequently passes by your house. Could it be that they are anxious and scared of something? Is your dog constantly barking because they think that it is playtime? Have they not been conditioned to stop barking long before this? If you look at the root of why they're barking so much, you might find a solution. For example, barking outside at neighbors walking by could be fixed if you get ticker blinds and block their view. If they're barking because they're anxious, you can do some things to alleviate those feelings.

Don't yell at them to stop barking. They will just think that you are barking with them. If they don't stop barking and all you do to stop it is add even more noise, the dogs will believe that what they're doing is right.

Wait for them to stop barking. It is frustrating but it is the best thing for them. After they've stopped barking, wait a few moments, and then give them a treat. You have to ignore them as they're barking or else they'll think about what they're doing is right. If you reward them when they're quiet, then this helps to make sure that they know quiet time is when you are the most positive.

Teaching them when to talk could actually help them be quieter. If you want to really control your dog's barking, start by teaching them to know how to 'speak.' To begin, wait for them to bark just once, and then give them a treat. Continue to do this and start to say 'speak' as they bark just as you did when you were training them to sit and to lay down. Rather than just barking whenever they want, it will help if they're trained so that they only speak on command.

This is an intermediate trick, so you probably won't really be able to do it right away, but it will still help with puppies since they can learn faster than older dogs. If training them to speak doesn't interest you, that's not the only way to have them stay quieter. Rather than yelling at them to stop, ask them to sit. When you ask them to sit, it snaps them into that mindset of needing to be more relaxed and quieter.

Remember that some dogs are just louder than others. There are certain devices that you can use if they bark a lot. You could purchase a noise maker that releases a high frequency pitch as they bark. The dog won't like the sound, so it might help them to stay quieter. You can use the clicker when they are good and click it as they are barking. It could help them to instantly snap into a calmer mindset.

Greeting Visitors and Walking Properly

As we mentioned earlier, when dogs meet new people, they will be highly excited. It is always fun to meet new people, and dogs don't have the kind of social anxiety that makes us as scared of other people! They will walk right up to anyone in most cases. The problem with this, however, is that not everyone likes dogs. Even those who are dog lovers aren't going to be as interested in having dogs jump all over them. It can be a bit too much when there is a big dog greeting you. The same goes for when you are walking down the street. A lot of people are going to be excited to see your dog and want to pet them, and your dog will notice their excitement, sometimes matching that and getting happy and jumping on them. There are a few things that can make this process easier.

Don't punish them for their enthusiasm. They are just happy to be meeting new people! They are simply excited that they are in a new environment and dogs aren't always the best at containing that kind of thrill! Rather than making them feel bad for being so happy, instead, make it your focus to simply remove them from the situation. Step back and make them sit or lay down. Once they have calmed down, give them a treat as a reward.

Continue to reward positive behavior. When they greet someone and immediately make sure to stay sitting, use positive reinforcement. Remind them that they are a good dog for not jumping on the other person.

Remember that this is a training period. You can tell your visitors this so that they are more understanding and they know that they might get caught in the cross-fire. For when visitors arrive at your home, keep the dog on a leash. Have them sit, and each time they get excited, make sure that they sit again until they are calm enough to have you let people come in. If someone knocks on the door and your dog starts getting excited and barking, make sure to not let the visitor in until they've calmed down. It is hard to not get your dog excited, especially if you are seeing someone for the first time after a while. You might want to tell them, "So and so is here!" so that they grow their excitement. Resist this so that they're as calm as possible when new people come in.

When you get home and your dog starts to jump on you, rather than stepping backwards, step forward. You want to remind them that you are in charge and you are the leader. They have to follow your lead, you aren't to follow theirs. Ignore them and don't greet them as they jump. Once they calmed down, this is when you can pet them, take them outside, give them a treat, and so on.

If they jump onto the couch and you want them down, touch the ground with a treat in your hand. Each time they jump up, do this, and have them sit on the ground. You can start to say to them, 'down,' so that they associate this instance with sitting peacefully on the floor. It will make it much easier for you and your family to tell them to not jump on them or on furniture when they have an actual command to get down.

As far as going for walks is handled, if someone comes up and asks if they can pet your dog, tell them that's fine, but make your dog sit down first. Remind the other person that they're in training and you don't want to teach them that jumping on strangers is acceptable. Having your dog know their name is going to be very important in this process as well!

Knowing Their Name and Coming on Command

Your dog's name is a very important part of their identity. This is their command. This is what they know themselves by. They associate this name with positivity, and it helps the two of you connect further.

Your dog knows to listen to their name because it means that you will be providing them with something afterwards. Calling them by their name from the beginning is important. Pick a name that is short. While Mr. Hashbrown Casserole the Third is an adorable name, you want something that is easy to call them. If you do have a long full name, that's fine, but have a nickname that they are most known for going by.

Some individuals even think giving your dog a human name will make it harder for you to treat them like a dog and not a human. This is up to you, but keep in mind that naming them after a family member or friend, or with a name that you already know might make it harder to differentiate your emotional evaluation of your dog versus this person.

Don't confuse them with a word that sounds like a command. For example, you might name them Brownie, which could sound like 'down.' Mitt might sound like 'sit' and so on. Their name should be clear and unique to them so that in all situations they always know who you are trying to refer to.

Something with a 'c' or a 'k' can help your dog to better understand if you are saying their name or something else. It is a noise that separates their name from any other noises that they might be hearing.

If you adopt them, there's a good chance they already have a name. Many breeders will refrain from name-training them. It is perfectly fine to change up their name (Keliher, n.d.). As long as you don't call them by their old name anymore, it shouldn't be an issue when training.

Looking at them and saying their name is enough. Make eye contact and say their name. give them a treat after they respond in the beginning. Don't overuse their name or else you might confuse them or teach them that they don't need to pay attention. If they aren't listening after the first few times, take a break and try again later.

As far as calling them, ensure that you start by saying the name around them. Give them a treat whenever they come to you after you've said their name. Don't make them "come here" when you are angry at them and want to use a negative tone. If you use negativity around their name, they are going to reject it and be far less likely to actually listen to you.

Don't use their name to call them when you are giving them a bath, taking them to the vet, or doing something else that they don't like. This just teaches them that they should be afraid of their name and of coming to you when called. If you aren't nice when they come to you, then they simply won't want to come to you anymore.

Fun Tricks for Your Dog

Some individuals might think it is tedious to teach dogs how to do fun tricks just for our entertainment. It is actually good for them too! It keeps them focused, and when they're obedient, they will be better behaved in all situations.

You can start with a simple 'shake.' This is a cute way for them to interact with new people. Since they do it while sitting down, they'll likely be more obedient as well. Have them sit down. You can simply wait until he puts his paw in your hand and then reward.

You could also try to tap the back of their leg to get them to put their hand up. Don't grab their paw, however. It is the same as trying to force their back down when teaching them to sit. They have to learn how to do this action on their own. It should be a natural motion. It will take some time and you have to be patient, but they will learn how to do it.

For rolling over, have them lie down. Hold a treat behind their backs and over their shoulder so that they have to roll over in order to sniff the treat. Get them to lie on their back, and then have them roll back into a lay down position. Once they completed the roll, give them a treat.

Use a treat to have them roll onto their back over and over again, never giving them the treat until they've completed the entire roll. Once they've done this, it becomes easier for them to understand what it is that they're doing which gave them the treat as a reward. Do these tricks daily in order to get them to understand the different steps they have to take to complete the trick for the reward.

Once they complete a roll, use verbal praise as well. Treats help of course, but it is your enthusiasm that is also going to make sure that they understand what they're doing is right. They will always be looking for your approval!

Having them "play dead" is super fun and will help them be the entertainment center of the party! They will love the attention. It is the same trick as rolling over, but instead of finishing the whole twist, they'll stop when lying on their back.

For this, you can associate a hand signal. Some people will hold their fingers out like they shot the dog, making it lay flat on its back. Others might ask a funny question, like, "Would you rather be a stinky cat or be dead?" This part will be up to you! It can be fun to play around with your command.

Reward them when they are in the position of playing dead. Use the treat in your hand when you make the hand gesture. This way, they'll know that the hand signal is what keeps them laying down.

Don't give them a treat right away or else they will think they have to snap out of this position quickly. Have them lie there for a moment and then give them the treat.

Practice these daily with your dog. They won't learn right away, and it will take some time for them to understand. Alternate between giving them a treat and doing it with no treat so that they do it for positive praise and not just a snack. If you give them a treat every single time they do it, and then just stop giving them treats altogether, they'll eventually stop doing the trick because it no longer means they get any treats. Don't give up with your dog, remain patient, and always remember positivity!

Chapter 6: Kinds of Exercises for Your Puppy

Playing with your dog and taking them out to the bathroom is important, but they also need to have specific moments of exercise. Just like humans, dogs can be more at risk when they aren't getting the healthy exercise needed. It could be bad for their joints, their hearts, and their bodies in general. Don't wait to exercise them! Start healthy habits from the beginning. You might discover that it actually helps you be more active as well!

Training and Agility Classes

A good method of exercise is to actually have them work with a personal trainer. If you have the means in your budget to hire someone to train your puppy, then this is going to benefit them and you. You can work alongside the trainer to learn important tricks as well as breed-specific training to help them be the best dogs they possibly can. Training isn't just for disobedient dogs – any of them could benefit! It is not even something you have to do right away either. Consider this as an option when deciding what kinds of exercises for your puppy to use.

Not everyone can afford this, so you can start to use training and agility cues at home. These are courses designed to test your dog's mental and physical skills. It is healthy for them because it provides them with exercises, and you have the ability to work with them directly, strengthening your bond. Your dog will be excited to complete tasks while also getting a fun workout that will improve his body health.

It is important that humans keep a critical mind so that they can function higher, but this isn't just true for us. Dogs with a critical mind will be more obedient. They will feel more connected to you. When they have more agility training, if you command them to do other things, they will be more likely to respond.

Agility is the kind of sport that dogs participate in where they run through hoops, fences, and other obstacles. You can purchase courses for them or make them on your own. Some individuals will use PVC pipes or pool noodles to create courses. There are common ones that you can try to recreate, or you could also come up with your own.

Your dog's safety is the most important part. Ensure any homemade courses are safe and that the dog's health and safety isn't at risk. It can help alleviate anxiety and make them more tired each day when you do consistent agility training. We just have to ensure we aren't pushing our dogs too hard when they might be feeling tired.

Start simply with some hurdles to jump over. These can be purchased at most athletic stores or you could use your own material. From there, you can set up cones for them to weave through.

To train them through these courses, you can use a simple treat in your hand to lead them through, or a longer stick with a treat at the end to help them go faster. After going through the course several times, a day for a couple of weeks, they will know how to run through with a simple command. There are many videos on YouTube that can help you deeper understanding how to train your dog for agility skills. It should be fun for them and time to play with you, not a moment where they feel stressed and scared of performing.

Swimming

Swimming will be a great exercise for your dog to participate in. It is easier on their joints and gives them something simple to do. It is a great way to cool them off in the summer as well. They will be happy to have a task, and this is another exercise that can strengthen their bond with you. They will trust you to care for them in this time.

A lot of dogs love swimming, but a lot of dogs also equally HATE it. It can cause anxiety and make them feel like they don't have control. It is a different sensation for them and not every dog can handle getting wet or feeling dirty.

A life jacket is a good idea for starting with certain breeds. Most will be fine, but there are some breeds, like bulldogs or pugs, that might sink right to the bottom when they are first in water.

Don't take your dog out to deep water where there might be dangerous currents. Keep it on the shore in the beginning to make sure that their safety is entirely protected.

Make sure that they aren't drinking the water they're swimming in. Lake water can sometimes carry certain bacteria harmful for your dog, and pool water has harmful chemicals for your puppy.

Book 3 - Training Your Puppy Step-By-Step

Introduce them slowly. Tossing them in the water can be traumatizing. They just might never go back to water again if you do this! Ease them into it so that they can be comfortable with the water around them.

Dogs don't like to be dirty. It is good to go for a walk with them where they can have their legs be wet while still being in control of their body. If you are wanting them to go swimming in a body of water, have them walk along the coast with you. Have treats with you so that you can give them positive reinforcement when they let the water touch their feet. Don't force them in just yet and guide them along where the land meets the water. Let them sniff and pat their paw in the water. They'll be curious so encourage that in them!

If you are taking them into a pool, buy a kiddie pool first and fill it up with just a few inches of water. Throw a ball in or a different toy they like so they can retrieve it and feel the water on their feet. Let them splash around in this safe area before you pull them into the actual pool.

Give them the option to leave if they are feeling uncomfortable. You should keep them on a leash since they might even run in fear but let the leash stay loose enough, so they have plenty of space to run away if they get scared.

Never command them to go into the water and only use praise and excitement. Don't tug them into it by their collar either. It should be a natural thing that they decide to do on their own.

Once they're in the water and floating, they'll naturally try to keep themselves afloat. Don't assume all dogs can swim right away! You'll have to be there to help them stay afloat as well. Support their stomach. So that they stay close to the top of the surface. If they start to sink and their face goes under, it might scare them to the point that they freak out and try to frantically swim.

Guide them to the edge or the surface if they are scared at any point. They trust you, so don't make them break that trust by forcing them to stay down under the water! Wait until after you've gotten out of the pool and rinsed off to give any treats.

Always supervise your dog around water. Even though they might be a great swimmer after a while, they could still slip into a pool or get lost in waves from natural bodies of water.

Catch

Catch is a great way to exercise your dog, one won't always require you to be running around just as fast. It is good to run with your dog, but it can also be exhausting! Playing catch gives you the chance to relax and enjoy the outdoors while your pup gets plenty of exercise.

A lot of dogs will naturally catch. It is in their biology to want to retrieve things for their leaders. They know that you enjoy the ball, therefore when you throw it, they'll chase after it.

Book 3 - Training Your Puppy Step-By-Step

Start with a toy that they will actually want to play with. Not all dogs are that interested in balls. If they prefer a rope or other rubber toy, throw this for them. After you've done this a few times, you could transition to a ball.

Reward them as soon as they start to play with the toy. Use verbal praise to remind them that you are excited and that what they're doing is right.

Put it on the ground in front of you and encourage them to play. This primes them so that when they catch the ball, they bring it back to you rather than run away with it. Once they bring it back to you, you can then give them another treat.

This is a good time to teach them to "drop it" too. Many dogs will bring the ball back, but they won't always let it go. They can't eat a treat with a ball in their mouth, so you can tell them to "drop it" and when they do, give them a treat. As they're eating you can take the ball from them. This is a great method to help them learn this trick. Next time they pick up something they're not supposed to as well, you can tell them to "drop it" and they'll know what this command is.

Dog-Led Walks

Walking can be so routine for dogs. It is a way that they go to the bathroom, get a reward, then go back inside. Most of the time, they take your lead. As a fun exercise to do with your dog, let them be the ones to decide which way you are going to go. Some individuals believe that you shouldn't let the dog lead because this is destroying your role as the alpha. That is debatable, so it is up to you to decide if this is a method you want to incorporate or not.

Every once in a while, you can still let them take the lead. Don't let them pull on it, but if they want to go to the left or the right, backwards or forwards, let them be the ones to decide. Keep a loose leash so that there isn't as much strain as they walk around.

This is best in nature areas rather than structured streets where they might be passing a bunch of other people. There will be fewer distractions so it will be more about them exploring their surroundings rather than getting excited about new people.

Go somewhere that you could take multiple paths. Give them options of how they can explore. It is fun and relaxing for your dog. It makes them feel natural and like they are accomplishing something.

Creating a Routine

The most important thing to do for your dog is to create a routine for it. Dogs need to be able to stick to a schedule. It gives them assurance that they'll be safe and protected.

Make sure that you are feeding at the same time every day. If they have a random feeding schedule, it will be stressful for them because they'll never really know whether or not they will have food to eat. They'll also worry that they won't be able to eat when they're hungry.

Book 3 - Training Your Puppy Step-By-Step

They will go to the bathroom around the same time as well. It should be a routine for them to wake up and go outside, go before bed, and a few similar timeframes throughout the day.

Sleep is essential for them. Some dogs can get anxious and not be able to fully sleep. If they don't have a schedule, they might not be able to become relaxed enough to get that deep sleep needed.

Most importantly, have an exercise routine. Not only will it be good for your pup, but you'll benefit from these periods of high activity as well. They will know the difference between play time and relax if you make sure that you are carving out specific times for them to exercise. Rather than having an active pup that acts all wild, you can have a dog that knows when it is time to lay down and when it is time to be crazy and play.

Make sure that you are also conscious of the methods that they might be associating place and specific locations with parts of their routine. For example, ensure that they have a proper place to sleep every night and that it is consistent. They will know when you take them to this place, that it is time to sleep. Have them eat in a similar place as well. If they are led to their food bowl, they'll know that it is time to eat.

Not every day is going to have the same schedule, but when you are doing those routine things, they should have the same structure each time. Keep procedures consistent so that there's no confusion on your dog's end.

Taking them to the bathroom first thing in the morning is a good habit. At the same time, it is ok to wait just a few minutes if you need to set up the coffee or go to the bathroom first yourself. You don't want to rush and stress your dog out with that first morning bathroom trip. Give yourself enough time so that this period can be calm and relaxing for your dog.

15 minutes a day is the minimum playtime you should have for ALL dogs. The bigger the dog, the more exercise they will need. If they have a day

where they don't get out much, make sure that they're twice as active the next day.

Chapter 7: What to Do for the Health of Your Dog

Your dog has plenty of health needs just as humans do. Choose a vet in your area that you trust who also has good reviews. Though you might be able to train and exercise them properly, it is their health that is still important at the end of the day. Ensure they live a long and happy life by always checking in with their health.

Initial Shots

It is important that we take care of our pet's health by giving them regular vet checkups. Going to the vet can be expensive, but it is something that needs to be done. If you can't afford regular vet check-ups, it is probably best to wait until you have some money saved up to adopt a dog. When you do get your puppy, you should have a savings account with around $250 - $1,000 in it for any pet emergencies. You don't want your dog to get a random sickness or a serious injury that you wouldn't be able to afford to help them out in the end!

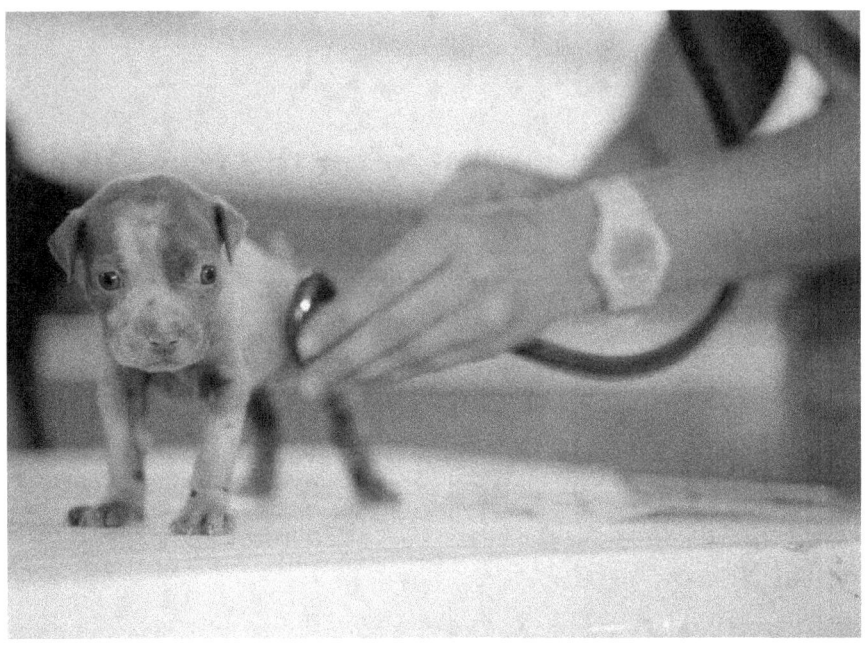

Many dogs will already have necessary shots if you adopt them from a shelter. This is a great option for those who can't afford all the shots up front. It is more convenient for you as well, so consider this before you adopt from a breeder.

Having vaccinations are also important so that when your dog is exposed to other animals, it can't catch, or spread any diseases. Puppies don't have the ability to get all the shots they need right away, so if your adult dog isn't vaccinated, it could affect your puppy. Aside from a few rare cases of having various reactions, getting your dog vaccinated isn't going to have any negative side effects.

Some laws require mandatory vaccinations. Check in with your local vet to see which ones you have to get in order to abide by your area's laws.

There are certain core vaccines that your dog will have to get. These are common health issues among dogs that you'll want to protect your pup from ever having to get. According to the ASPCA, these are the illnesses and diseases that core vaccinations will prevent:

- Canine parvovirus
- Distemper
- Canine hepatitis
- Rabies

There are other vaccinations that might be necessary for your dog. These could be things such as:

- Bordetella bronchiseptica
- Borrelia burgdorferi
- Leptospira bacteria

For puppies, the vaccinations should occur within six to eight weeks of their life. They will only get a few at a time with three-week periods in between (ASPCA, n.d.). By 6 months, they should have all of their proper vaccinations. Check-in with your vet to see if there are any that they'll have to come back for as they age.

It will be up to your vet and your dog's needs to determine what vaccinations are required. Its best to do as much as possible because the only downside is that they might cost more money. These vaccinations will often be a one-time thing, so it is not like you'll have to continually save up more money to keep up with regularly purchasing them.

Don't administer vaccinations on training days. They're going to be tired, and likely stressed out from going to the vet already! Give your dog a chance to rest and heal. The biggest issue you'll notice is that they are tired and that the injection site might be tender.

Anything too concerning, contact your vet. For the most part, however, they'll be completely fine and safe from getting anything serious.

Spay and Neutering

Not everyone will choose to spay and neuter their dog. You might use them to hunt or breed, so instead you'd prefer to keep all their reproductive systems intact. This is up to you, but unless you need their reproductive organs for a specific reason, you should be getting them spayed or neutered.

Spaying is when female dog's organs in their reproductive system are removed. Neutering is for males. For females, the ovaries and uterus are typically removed. For males, their testicles are removed. The actual penis and vulva are not removed on dogs. Emotionally speaking, it is a hard process for your dog to undergo, but in the end, they will be happier and healthier because of it. They might have a more relaxed temperament, gain weight easier, and be calmer in general.

It is important first to ensure that they don't reproduce. Having one puppy is hard enough, so you don't want to have more to take care of at once! After around six months, they will be able to start reproducing, so this is the age that they're usually spayed or neutered. Some studies have actually shown that waiting a year to spay or neuter larger breeds, like German Shepherds, is actually better for their health to give them the full chance to develop first (Dayton, 2016). Whatever you do, try to spay or neuter your dog between six months – 18 months of their life.

For female dogs, you will cut out their heat cycle. Some female dogs might even bleed when they are going through their heat cycles. It doesn't do much to them other than cause uncomfortability. They will likely be licking that area more when they're in heat, so they could end up causing sores if they do it too frequently. The bleeding could happen throughout the day as well, so it could end up staining your furniture or wherever else the dog might sit.

For male dogs, they might resist the urge to 'hump' things. Unfixed male dogs will usually do this to other animals as well as many toys. Even after they're fixed they might still do this, but it will be less frequent. It is usually not a reproductive issue and something that you have to train them to stop doing. Catch your dog before they start to hump. As you see them mounting something, command them to sit or lay down. Once they've successfully done this, reward them with a treat.

Both female and male dogs will also be more likely to listen, and it will be easier to train them. They will come on command easier and you won't have to worry about them being disobedient.

It can drastically reduce the risk of certain cancers or infections, such as pyometra. The lifespan of dogs that are fixed versus those that aren't is also longer.

You might be weary because you worry that it is not a natural process. It is also not natural to deprive them of these urges. It can end up hurting them in the end. They will still seek out other dogs to reproduce with and they might get stressed because they aren't able to find a mate. If they are fixed they will naturally be calmer, and they won't think about this as frequently.

Keep them away from other animals once you have gotten them fixed. They might be extra sensitive so you don't want to cause a fight that could make their stitches become loose.

Reduce their play time for the first few weeks after getting them fixed as well. You don't want to push them to overdo it or else they might end up hurting themselves in the process.

Give them two weeks to recover. This is the usual time period it takes for them to fully heal. They are getting internal surgery, so it can be an exhausting act to recover from.

A cone is probably going to be necessary because they'll want to lick their wounds. Make sure to keep this on them at all times that they are unsupervised, and even when you are together. It could only take one lick or bite to remove stitches, which could mean having to go back to the vet and do the whole stitching process over again.

Picking the Right Food and Treats

As you can see by now, the best method of training your dog is to use positive reinforcement and lots and lots of treats. For both food and treats, there are a few health guidelines that you should try to follow. Dogs aren't like us in that they won't care about tasty food as much. We have over 9,000 taste buds while dogs have far less, estimated at 1,700. They can still taste, it is just not as important to them. Many humans don't realize this and will often put their human emotions onto dogs. They think that these animals deserve to eat complex meals just as we enjoy, but this doesn't matter as much. We should be providing them with good and healthy food, but they aren't going to need to eat a steak dinner every time you do, so don't feel guilty that their food isn't as tasty. It is the smell of this food that they will care about more than anything (Finlay, 2017).

Book 3 - Training Your Puppy Step-By-Step

Dogs are also omnivores. Unlike cats, they should have some fruits or veggies in their diet. Raw veggies, like carrots, frozen peas, bananas, and apples are all great treat substitutes. Since you will be training them so frequently in the beginning and providing them with a ton of treats, you will want to consider these natural options to keep their weight under control.

The food you dog consumes will directly affect how they behave and how long they might live. We want our furry friends as healthy as possible so they can live a long time by our side! Choose the right food with added benefits and natural substances in order to ensure their health is in tip top shape.

Foods that are cheap and the popular brands that you can find at any convenience store will more than likely not be the first choice. Many dried foods are filled with carbohydrates that just add calories. Yes, dogs have to watch their carbs and calories too! Don't feed your dog only junk food. Many popular brands are equivalent to humans eating chips and other junk food as their sole meal.

The first three ingredients are the most important. Check any bag of food and look at the things that are listed before everything else. If these are chemicals that you can't even read, you might not want to have this be something that you give your dog.

The worst ingredients include:

- Meat by products (typically carcass or other animal parts that aren't actually meat)
- Caramel color (no dyes or additives should be in dog food)
- Oil (oil is dangerous for dogs and can cause inflammation)
- BHA or BHT (preservatives dangerous to health)
- Animal fat

The best ingredients will be:

- Meat
- Grains
- Vegetables
- Vitamins, minerals, and nutrients
- Meat meal

Your dog should be eating foods that have AAFCO approval. This means that it has actually been tested on dogs. Not all brands will give their food to dogs before even releasing it. Though it might taste good, it might not have any actual benefits for them.

Puppy specific food should be used for younger dogs. They have different nutritional needs that have to be attended to as they are still growing and developing. It will often be softer as well. Consider their breed, too. Some small dogs will have different food they need to eat, and larger dogs might have various needs as well. The tenderness and the size is important.

Wet food may be a better choice. The food that they eat in the wild is wet and moist, not crunchy bits all the time; however, you might like dry food because they don't eat it all at once. Check with the package to see what the serving size is. For puppies you will want to feed them three or four times a day. Once they're a year old, cut it back to just twice a day.

Water should be out all day long. If you have multiple pets, you should have multiple water bowls. It will be a good idea to have one more water bowl than people that you have. Never limit their access to water and ensure that no matter where you go, they will have the chance to drink as much as they want.

Avoiding People Food

It is hard to not give your dog a little snack when you are eating. You want them to like you and when you are enjoying food, you might want them to get the chance to enjoy this just as well! However, if you are giving them too many table scraps, it will create bad habits.

Book 3 - Training Your Puppy Step-By-Step

Not all people food is harmful to dogs, but you should avoid giving them your food when you are eating. It will create negative boundaries with your dog. It is a good idea to keep them in their kennel when you are eating so that they aren't trying to bother you too much while you have food in front of you. You should feed them after you eat, as this indicates to them that you are the leader of the pack, since in the wild, the leaders get to eat first.

Avoid giving them any human meat because there's so many factors that could be harmful. Even if you get it at a restaurant and want to give them scraps, you never know what they might have seasoned it with initially.

Fruits and veggies are fine snacks. Give them to your puppy as you wish but remember to not do it in the space that you eat. They should know to respect your boundaries and leave you alone as you try to eat your meals. These healthy fruits and veggies include:

- Carrots
- Bananas
- Apples
- Watermelon
- Cucumbers
- Peas
- Sweet potatoes
- Celery

These are great treat alternatives. If you don't want to buy a ton of treats all the time or find that they cause a ton of weight gain, then switching to these foods is a better option.

Make sure that anything you give them isn't seasoned. Dogs can't have garlic and onion, among other flavors, so you want to ensure that there are no hidden seasonings on the snacks you might be giving them.

Make sure that you never give them scraps directly at the table. They will know that it is fine to beg. Take the scraps to where they eat. Create a habit of making them sit or lay down in a different room when you are feeding them. You might think it is fine to do this at your home, but when you go to a friend's home or have entertaining guests over some meals, you don't want your dog to constantly bug them.

Ask yourself if your dog would find it out in the wild to eat on their own. They might naturally find certain fruits and veggies, but popcorn, chips, and bread they wouldn't. The key is to give them things that are going to help their health, not just satisfy their taste buds. If your dog doesn't eat their dog food, then you'll have to consider if they simply don't like it and purchase a new brand.

Flea Treatments and Bathing

Fleas are annoying, but they won't be the worst thing that can happen and are usually easily treatable. The worst thing that you will find if you get fleas in your home is that you get bitten and your bites might be itchy and swollen. This is treatable with some Neosporin ointment or other anti-itch cream to help your bug bites. It is worse for your animals, however, as the fleas can actually feed, and breed off them. Not only will the fleas bug your pets, but they will scratch, and this could cause even bigger sores and tender spots on their bodies.

They will get trapped in their fur and it can cause your dog to aggressively itch themselves. We scratch our wounds as humans, but we usually know when we need to stop before it gets bloody. Your dog doesn't know as well and will scratch and scratch and scratch. This can make their wounds even bigger and cause infections. If you notice your dog frequently scratching, it might be a sign that they do have some sort of insect issue.

Though fleas are more treatable than other issues, don't pretend like they're no big deal, because they can still carry viruses. Some fleas can go from two to 2,000 within a couple of weeks!

You can look for fleas yourself by separating their fur and looking at the root of the hair. On their skin you will be able to see the flea crawling around, or you will notice the flea dirt.

When you do see a flea, the only way to ensure it dies is to drown it, as squishing them doesn't always work. They have a hard-outer shell, so it is often the best choice to flush them or rinse them down the drain with soap and rubbing alcohol. If you do find a bug and it is not very tough to squish and kill, then it is probably not actually a flea and instead some other sort of insect.

You can give your dog a flea bath and then give it a flea treatment. The first bath will be to make sure that as many fleas and flea dirt is washed away as possible. You'll then want to give them a flea treatment. The best kind is an ointment you put on the back of their neck. This soaks into their natural bodily oils and will eventually spread throughout their entire body. About a half a week to an entire seven days from giving the flea treatment, you could do another bath to get rid of any fleas that might have died. Check the box of the flea treatment you purchase to see if you need to wait longer.

Treat your dog not just for its own sake but for other pets around the house as well. Your dog that goes outdoors might bring home fleas to your indoor cats. If you have indoor cats and a dog, you will still want to give them treatments. The dog might spread fleas to them, and then the fleas could spread back to your dogs. It is best if simply all cats have been treated.

If your home seems to have fleas pretty bad, you can do a home treatment. This is usually in the form of a smoke bomb that will spread throughout the entire house. Ensure that no animals or humans are home when you do these treatments.

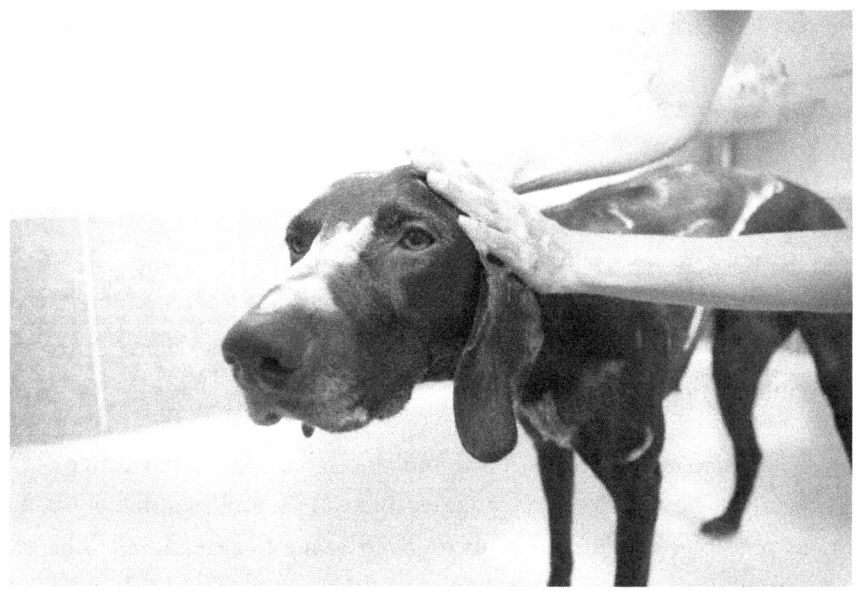

As far as regular bathing goes, this will vary depending on the dog. Some dogs will need regular bathing while other dogs will be perfectly fine going a year or more without a bath! It will depend on the dog breed whether or not it needs bathed.

Short-haired dogs will need bathed less. Make sure that you regularly brush all dogs, short or long haired. They will still need brushed in order to remove dead skin and loose hairs. It will help regulate the hair they leave behind on your furniture as well. Outdoor dogs that are very active need more bathing as well.

Once a season, or around three-or-four-month intervals is usually the minimum for bathing. This would be for bigger short-hair dogs who don't need as much grooming. The most frequent you should do is every other week. This would be for dogs that need regular grooming and whose hair grows long.

Keep their hair groomed so that it doesn't get too long between their toes or too long over their eyes. If you live in a hot area, it is best to maintain a buzz cut so that they don't get too hot. When it is cold outside, having a thick coat will help them stay warmer and make it easier to encourage them to go outside.

When bathing, pick only a dog-shampoo. Sometimes people will use baby soap, but there might still be dyes or perfume in this that could be fine for a human but not for a dog. Make the water warm, and have the environment be relaxing. Try to ease your puppy into the bathing process by making it exciting and giving them treats. Some people have actually used the trick of spreading peanut butter on the bathtub wall so that the dog licks it while they're getting bathed!

Chapter 8: What Not to Do When Training Dogs

There are some common mistakes in dog training that you'll want to do your best to avoid. There are many dog owners in the world that might not give their puppies the best chance when training, and then they end up wondering why their dog isn't behaving the way that they want it to. To make sure that your dog is going to be properly trained, we've created a list of what not to do so that you don't make any of the common mistakes that other pet parents might!

Aggression in Training

As we already discussed, aggression will just make them react more aggressively in the end. Though they might get frustrating at times when they don't listen, and they might have times where they don't have any progress in training, using aggression on them can cause them to be afraid and will actually regress any valuable training you did use. These are the most common forms of aggression and you shouldn't be using it on your dogs ever:

- Hitting
- Kicking
- Forceful pushing
- Threatening body stance
- Verbal aggression
- Yelling
- Pointing
- Other forms of physical and verbal violence

Book 3 - Training Your Puppy Step-By-Step

Some dogs are rather aggressive, so you might ask yourself, "What am I supposed to do to control my dog?" You might have a 120-pound dog that never seems to listen when you are telling it to sit down. Perhaps you have a little poodle that never stops using the bathroom in the house. Though these are frustrating situations, violence isn't going to help. We're going to discuss a few ways that you can assert power that you don't need to use any violence on them.

Not only does hitting your dog hurt them physically, but it destroys that bond. You are the one person that they trust more than anyone. How are they supposed to trust you if you hurt them so much? They won't be able to listen to you because they're scared of you. The more aggressive and angrier you are with them, the more that they will have trouble following your lead. They will grow their fear, not their trust, and that's a quick way to kill anything the two of you have built together.

Let's first understand the reason that they're aggressive. Dogs are very complex creatures and they have some needs that have to be met, while also having different stressors and anxieties just as humans do. Though they don't feel the same way that we do, they can still experience some of the same effects of stress.

The first reason is because they are trying to be protective. Dogs are territorial animals, like many other creatures. They want to be able to help protect your home. They care about you and want to ensure that you are protected. If they are often aggressive towards others, they are likely trying to protect their turf.

They are often frustrated and feel the need to gain control in some way. Just like how some people are more easily flustered, many dogs can be confused and turn angry too. They might be overwhelmed by others and feel the need to be aggressive to try and gain control. When you don't know what's going on, you might find that you panic, and this is how some dogs will operate.

They might be aggressive because they haven't been socialized. This is why it is important that we start to socialize our puppies from the moment that they're brought home. If we don't, we could end up making them scared of visitors and more likely to lash out on others.

Animals have a fight or flight instinct just like humans, so showing signs of being aggressive might simply be because they are anxious. In order to assert dominance, there are a few things you can do.

Be consistent with the boundaries you set. You can't have rules that apply during one part of the day, but not the other. You might not care if they jump up on you, but you have to set this boundary so that they don't jump up on strangers.

Sometimes, forcing them to be affectionate can actually damage that bond. If you are constantly squeezing and kissing them, or even holding larger dogs that don't like to be picked up, it could make them more fearful. They might not know what you are doing, so let the dog come to you when you want to show your affection. Be gentle and patient with them. If they are sitting next to you and licking you, you could try to put your arm around them and be affectionate, but if they're running around, eating, or playing, don't try to force a hug.

You have to have confidence over them. They will smell fear. If you are stressed, it might make you stress more, and then you could end up releasing chemicals like adrenaline through your glands. Dogs can smell this and know that something is wrong. If you are frustrated and stressed out when training them, they will sense it and it will be harder for them to focus.

Never stop playing with them! Get their energy out through playtime, not through times when they are rough and constantly jumping on others.

Feed them after you eat your own meal. As we already discussed, this is common for the leader of the pack. If you want your dog to really listen to you and remain obedient to what you are asking them, you have to ensure you are establishing dominance over them in a way they respect. You can also take them for a walk right before they eat so that they know you are in charge. You are making them work for their food!

Inconsistencies

The biggest killer for your dog's training is not being consistent throughout the process. Dogs need routines and they only know what boundaries are in place by experiencing them on a repeated basis. If you are consistently inconsistent with your dog and the rules are always changing, they won't know what they should be doing.

What you do at home is what you have to do everywhere else. Even if it might be hard to remember, always keep your dog in training. It will be a long process that will take time to get used to. They need to follow certain boundaries in order to stick to the rules which you are teaching them.

Their behavior should be right or wrong at the core. It is wrong to jump up on people. They won't know that it is wrong to jump up on strangers, but okay to jump up on you. They will simply know that it is fine if they do jump.

Even when you don't feel like doing something, you have to stay consistent for your dog. It can be hard to always take them out at the same time, and there are moments where you might just want to stay in bed a little longer even though your dog's already woken you up to go to the bathroom. Even in these moments of struggle, it is essential that you are taking care of your dog's needs and staying stable.

Different partners should not have different rules. This is why it is so important to have a family meeting to establish all of the rules. Even if you have a roommate who's not as interested in caring for the dog, they have to help out just to ensure that no bad habits are created.

Dogs don't have a way to communicate with us other than through signs or symbols. They won't be able to see these if they aren't consistent and presently clear. They need that regulation to understand, as it is really the only true way that they can communicate with you.

Useless Repetition

Your dog needs routine and consistency. However, if you are consistently doing something that's bad for your dog or that they don't understand, then this isn't an ideal routine that will help them in the end.

Useless repetition is like calling their name over and over when they clearly aren't coming. It is like trying to make them sit by pushing down their butts over and over, never able to actually do it on their own. When you notice that you are repeating something to your dog over and over again and they don't get it, then it can be frustrating. However, there are certain steps that you'll want to take instead to get them to really understand what you are trying to teach.

For example, you might find that you are trying to teach them to roll over. You've done it twenty times now and they still don't get it. The first thing that you'll want to do is ensure that you take a break. They might simply be tired, and you could be showing frustration, a bad recipe for pointless training.

Your emotion will change throughout the useless repetition. At first you might be excited with a happy tone while you are training your dog. By the end, your voice might shift and you could end up sounding frustrated. This will only confuse the dog further as they won't really understand what it is that they did wrong to make you frustrated.

Grabbing his paw and shaking it over and over again isn't going to magically work on the 50th time when you are trying to teach him to shake his paw. When something isn't working, you have to reevaluate the process to find a method that they actually do want to participate in.

Sometimes, the repetition is the way that they end up learning. They think that you have to get angry and say 'sit' five times over and over again until they actually decide to sit. Remember that your dog won't think like you do, so they aren't going to understand the pattern you want if you are trying to teach it in an unhelpful way.

Confusing the Dog

It'd be so much easier to train dogs if we knew how to really talk to them. Unfortunately, human-dog communication is still limited and rather than finding methods to speak to them, we have to discover the way that they will understand our signals. Let's discuss what you might be doing wrong which could confuse them. If the dog doesn't understand what you are trying to do, it can make them upset and then it will be harder to train them in the end.

The first thing that confuses them is when you punish them for natural things. This includes things like going to the bathroom, eating, and playing. If you notice that they are going to the bathroom in the house and scream and yell at them, they will think that they're getting in trouble for simply going to the bathroom. If you get home and see that they are eating food you had sitting on the counter, they will think that they are in trouble for eating. You have to catch their bad behavior before it happens, or right as it starts. Punishing them in the middle of it will simply be confusing you.

Did they step on your foot and you yelled? You are not likely to be able to train your dog to not make mistakes. They will still stand behind you when you are cooking sometimes, or they might jump onto your bed and accidentally stick a paw in your gut. Though these accidents are annoying, you can't punish your dog for this because they won't really understand what it is that they're getting in trouble for.

Did they have diarrhea on the floor? They're sick, not intentionally being bad. Though it is bad for them to go inside, they won't be able to hold it all the time, especially if they are ill. The same can be said about them vomiting. Though they might be potty-trained, there are still some accidents that are entirely out of you and your dog's control.

Having the wrong labels for certain cues can make it hard for them to know what to do. Check and ensure that you are not training them to do a certain command with several different words. For example, you might use "sit," "down," or "sit down" all just for the same command for them to sit. Though they all mean the same thing, your dog doesn't know this, and they could end up getting confused.

Try to refrain from using your finger and hand to "punish" them. They will really only understand your use of different words and symbols, not necessarily hand gestures.

Make sure that the affection you are giving them is at the right time. Even though you might want to hold and snuggle them all day long, if they don't seem into it, you can't force it. Your dog might end up thinking that this is some form of punishment, but they won't be sure why you are punishing them in a certain way.

Pay attention to the eye contact that you are giving your dog. They take a lot into consideration when evaluating your eye contact. If you are staring for too long it could make them nervous or anxious. They might even end up getting aggressive depending on the context, thinking that you could be challenging them.

Abandoning a Ritual or Habit

Dogs will need trained throughout their entire puppy-life, and into adulthood. You can't stop training your puppy just because they seem to know the same old tricks. As they get older, you won't have to train with them so often, but there are still important measures to take to ensure they keep up with their knowledge on certain tricks. You might train your dog to play dead within the first few months of their life. If you stop doing this trick, by the time they're two years old, you can't expect them to immediately remember how to do it.

Just because you started once doesn't mean that they will know this forever. In order to ensure that they are keeping up with the tricks that you taught them, have moments where they have to do the treat in order to get a certain reward. The best time to give them trick refreshers is right before they eat, before you take them outside, or after you come back in from taking them out. Give them a treat and positive reinforcement. It only takes them doing these tricks once every couple of days for them to maintain their knowledge of a certain command.

Sometimes, you might abandon an entire ritual. Perhaps you no longer take them outside first thing in the morning and instead wait until it is convenient for you. This can mess up your dog's entire schedule and make them rather confused! If you do need to change protocol, remember to transition your dog. They are already used to a certain way of life and become easily confused. The sudden change could stress them out, so do your best to keep your dog as calm as possible.

Treating All Dogs, the Same

All dogs are different. Even though they have similar needs and will react in usual ways, they're also very different from one another. Some dogs are lazier, some are more aggressive, some are very needy, and others might be independent. Whatever you might have thought about dogs before could be completely wrong after meeting that one special dog that just isn't like any others!

Just because you trained three dogs already doesn't mean this fourth is going to be the same. A mistake that some will make is thinking that all dogs act the same way. They might believe that if a dog isn't acting in the same way that another dog has in the past, that this new dog has something wrong with it. You might have just had a dog that was easier to train first, so the second dog isn't necessarily bad, it is just a little more difficult this time around.

Some dogs will be naturally shyer than others, while some may be boisterous and high energy. Others might simply be ornerier. Some dogs are excited and active, others are scared and want to sit away from everything. Some dogs love new people, some dogs get scared of strangers. Some dogs bark at cars, some dogs like riding in them. Some dogs will play with 100 different toys, some dogs might only want to play with one. Some dogs will be excited to see you, other dogs might not care as much. Each and every dog is different. Not only might they be born that way, but how they were raised before meeting you will also affect their development.

Don't punish your dog because it is not the same as others that you've had. Respect their individualities just as we do for humans! You can assume that some things are going to be similar, especially if that helps in your training. At the same time, remember that some of the things about their personality aren't bad, just different!

Positive Reinforcement with Bad Behavior

One of the biggest mistakes that new trainers can make is that they will end up using positive reinforcement for certain bad behaviors that their dogs have. This occurs when a dog is doing something bad, and the owner gives them a treat in order to stop this bad behavior. In the process, however, the dog learns that if they do this bad thing, they'll get a treat in return.

Sometimes you just want to get your dog to shut up! You just want them to stop barking, maybe they're scratching, or simply being violent. While giving them a treat is a great way to stop them, it also tells them that if they just bark enough, if they get aggressive, or if they start to chew something up, they'll get a reward in return.

Don't use a reward to get them to stop doing something. Instead, use a common command to get them to stop, such as 'sit' or "lay down." Once they have done this good thing, then give them the treat. Next time the dog is barking at cars and you can't get them to stop, don't call them over to give them a treat. Call them over, make them lay down, and then give them the treat. They will better understand that the only way to receive a reward is to be calm and relaxed.

Your dog is smart enough to know how to get what they want. This is just as we mentioned previously with crating the puppy. If they whine and then you let them out, they will know that whining means they can get their way.

Positive reinforcement in negative situations might confuse them as well. For example, let's say that your dog is really afraid of thunder. Whenever there's a storm, they go and hide under a desk. Every time you go to him and say, "it is ok, don't be scared," over and over while you pet him. The petting can help, but they will also pick up on your words. Next time that something scary happens and you tell them, "it is ok, don't be scared," they might be brought back to those same frightened feelings that they had the first time.

Not Proofing Tricks

If you train your dog to sit in the living room within the first three days, that's wonderful! The next step is to 'proof' this behavior. This means that you are taking a trick that they learned and doing it in different settings with various distractions to make sure that they understand the trick at the core. If you teach them a trick in one room, they might only thing that they should be doing that trick in that room. You have to repeatedly do this everywhere so that they understand what the command means at the core.

Book 3 - Training Your Puppy Step-By-Step

Dogs don't have the ability to make generalizations like we do. They're not going to understand that just because they learned to 'sit' in the kitchen they're supposed to sit in the living room as well. To really make sure that they're understanding this, you have to proof your tricks.

Start to do the same tricks in different areas throughout your home. Make sure that you do it in the backyard, on walks, and in all rooms.

Add in other distractions after this as well. Do it at the dog park where others are watching. Do it when you have their food prepped and ready to give them in their bowl. Create methods of excitement so that they understand they should be present in all moments and focusing on you.

Do it without a treat in between sessions. You don't want them to stop doing the trick just because you stopped giving them treats. You might have to teach your dog the trick over again, but this is the chance for you to strengthen their abilities. Patience is going to be key! Don't forget to proof or else your hard training work will be lost. If you are giving them treats and showering them with positivity, then they are understanding that this anxiety is fine to have in these scenarios. While it is not a bad thing for them to be scared, don't use positive reinforcement or else they will react the same way next time.

The best thing you can do to comfort your dog is to distract them. Get their favorite toy and play with them. Give them commands and tricks, rewarding that behavior with treats. Act normally and continue to be excited, not sad or scared. Don't force them either. Sometimes it is best to leave them alone while still being in the same room just so that they know you are there to keep them safe. You might put your arm around them or pet them, but refrain from using positive language or treats because they will get the idea this is how they should act in fearful situations, making them only more scared in the end.

Waiting Too Long

If you wait too long to start to train your dog, you are wasting precious time that could be spent helping them grow. Old dogs can learn new tricks, but it is going to be so much harder. Those who wait might grow frustrated with their pups easily and think that it is the dog's fault they aren't learning fast enough.

Start to train them from the moment you move them in. Rather than associating the different tricks with just you, they will associate this new behavior with their home. It is a great fresh start for them, especially when adopted, so take this opportunity to create good habits right away.

Be consistent! You can't get mad at your two-year-old dog that it can't walk on a leash still if you never trained it to properly use a lead.

This is the same for those that wait too long to 'punish' their dog. If you come home to find the garbage has been torn through, they're going to be excited to see you because they probably forgot about this. You might get upset at them right away, and then they will think that you are simply upset to see them.

Though dog-shaming videos and memes are cute, dogs aren't reacting this way because they feel remorse over what they did. The reaction of the sad face and low head are only because they see that you are upset, and they're scared. No matter how guilty they might appear to be, they are really just making that expression because they are afraid of your mood (Manning, 2014).

Book 3 - Training Your Puppy Step-By-Step

Chapter 9: Your Step-by-Step Training Plan

Congratulations on making it to this part of the book! You are almost ready to bring your puppy home. The most important part is that you create a schedule. We have a how-to on the way that you can do this, along with a few other key principles to remember before we finish!

Get to Know Your Dog

As we already mentioned, all dogs are different. When you first bring your dog home, make sure that you are doing your best to really get to know them. They will have their own fears, strengths, weaknesses, and interests. Some dogs will be the type that always want to be outside, and others will be perfectly fine with sitting on the couch and sleeping all day. While we did our best to ensure this would be a comprehensive guide for all new pet owners to know how to train their dogs, you should still look into further research to discover the methods that you could use for specific breeds. Though it doesn't matter as much what random facts you might know about your dog, knowing their history, their intended breeding purpose, and so on can really help you better understand why your dog might operate in the way that it does.

Create a Schedule

It is important that you keep your dog on a schedule, just as you would a baby or a toddler. It is not only a good way for you to get into better habits, but it is the method that your dog is going to be able to learn how to function properly as well.

You should start to feed your puppy four times a day until they are around three months old. The best time to do it is when they first wake up and a couple of hours before bed. Then, you can do it twice in between whenever it is convenient for you, so long as it has been spaced out.

Next, until they are six months, drop down to three times a day. The same rules apply when you are figuring out when to feed them. It should be at times that are convenient for you in order to ensure that it is a schedule you can keep up with. For example, if you are usually heavy into work or studying around noon every day, don't try to disrupt your schedule with a feeding here. Though it might be a better time for your pup, if it is not a good time for you it might get forgotten or overlooked, resulting in inconsistencies which confuse your puppy.

Once you get them fixed, you can cut it down to twice a day. Make sure that you adhere to serving suggestions on the packaging of your selected food so as to not overfeed them. eight hours is usually a good time to feed them. After they have eaten, you should take them out around two hours later, though when training puppies, every hour is ideal if you can. They have smaller bladders and aren't going to be able to hold it. As they grow older and learn to control their bladder movements, you can reduce these time periods so that they learn how to resist the urge to go when they have to.

Here are all the things that you will want to include in your schedule for a puppy under six months:

- Three to four feedings
- Five potty breaks
- Training session
- Exercise routine
- Playtime
- Bed time

Again, it will really be dependent on your specific and individual schedule when these times might be. Stick to what works best for you and always consider your dog's needs. Here's an example of what a specific schedule might look like:

- 8 AM: potty break
- 8:30 AM: feed
- 10:30 AM: potty break
- 11 AM: training session
- 1 PM: feed
- 1:30 – 1:45: crate
- 2:00 PM: playtime
- 3 PM: potty break
- 3:30 PM – 3:45: crate
- 4 PM: potty break
- 5 PM: feed
- 7 PM: potty break + 30 minute walk
- 8 PM: feed
- 9 PM: playtime
- 10 PM: potty break
- 10:15 PM: bed time

This can seem like a lot, but feedings and potty breaks will be reduced as your puppy grows. Play around with your schedule at first, but after about a week of being in their new home, your puppy should have a set schedule that they can depend on.

Keep it Simple

Don't overly confuse your dogs. Always look for ways to keep things simple, short, and sweet because this is the method that they are going to be able to learn from the best. If you discover that they are having trouble learning a new trick, it could be that you are making it too complicated. If you find that it is too hard to keep them on a schedule, it might be something too complex for you. Though this book has been comprehensive and long, it is also not as hard as it might seem to keep your dog on a schedule and properly trained. With effort and some troubleshooting, you'll be able to find the perfect routine for you and your furry friend.

Work on Rewards

Remember that positive reinforcement is the center of proper training. When you are able to provide them with rewards they really like, you'll have exceptional results. Experiment with different treats to find the ones they like the most. The more they like a treat, the more likely they are to respond when you are training them! Rewards aren't just in the form of treats either. They're also the positive words and pets that you use for them. Notice the spots that they like scratched or pet the most. Do they prefer their belly rubbed? Their ears massaged? Their chins scratched? When you know your dog's favorite rewards, you can use these to get them to act the way that you want.

Repeat, Repeat, Repeat

Just because your dog successfully does a trick one time doesn't mean that they already know how to do this. They will still need repetition to ensure that they always really understand the trick you are trying to get them to do. though it can seem tedious and at times exhausting, go over and over and over the same trick to make sure that your dog has been able to properly learn it. Like we mentioned earlier, useless repetition will only set them back in their training, so make sure that you aren't repeating something that isn't working.

Check-In with Your Dog

Always consider your dog's needs first and foremost. Get to know their emotions. What do they do when they're anxious? How do they react when they're scared? Are they getting enough attention? Are you showing enough affection? Though it might seem easy to really understand a dog, they are more complex than you'd think! Just because something works

Book 3 - Training Your Puppy Step-By-Step

for you doesn't mean it does for your dog. When things aren't going right, ensure that they are happy and taken care of. You are their caretaker, so you will always want to do regular check-ups with them to make sure you are providing the best care possible!

Conclusion

Don't feel nervous about getting a puppy, it should be exciting! At the same time, remember that it is crucial you train your puppy in a responsible way and never overlook the important skills they need to be a healthy dog.

Start by bringing home a dog that will work for you in the first place. Though you might like the look of a certain breed, an Australian Shepherd isn't good for someone who doesn't go outside often and enjoys sitting on the couch. Vice versa, if you are someone who enjoys an active lifestyle activities such as exercising, hiking, and so on, a mini poodle that sleeps for 20 hours a day isn't the top choice.

After you've picked your dog, it is time to get your home ready. Pick everything up that could be dangerous, and provide them with a safe space that they don't have to feel uncomfortable in. Give them necessary items and pick out the toys that will help them grow and learn. Give them plenty of treats and water, as well as carefully selected food.

Ensure that you start training from the moment that you get home. Teach them how to sit, and how to go to the bathroom outside right away. Every other trick that you teach them will likely build from sitting down first, and this is a great way that you can make them calm and collected whenever you need them to act that way.

Socialize them from an early age as well. They'll need to get used to controlling their excitement in public, and it is essential that we teach them good behavior with other dogs and people from the very start. This is especially important for aggressive dogs. You have to also ensure you are getting them out of the house to tire them out. Active and ornery dogs may go a bit stir-crazy when you don't get them outside and playing around enough.

Once you have a healthy routine and can start to socialize them properly, it is time to add in more challenging skills. These are things like rolling over or laying down. Once they are able to learn one trick, it will be easier for them to learn even more.

After all of this, ensure that you are giving them the right amount of exercise. Though you might be playing with them and training them at the same time, they will need separate moments of exercise. This could be in the form of agility training or going for basic walks.

At the same time, you'll need to ensure you take them to the vet for regular check-ups and that all of their health needs are being met. Give them proper baths and grooming and don't forget about their need for flea or tick treatments as well.

Never forget the common mistakes that other dog owners might make. Each dog is unique and needs to be treated this way, so even though something might work for one dog, that doesn't mean it will for another.

Your dog is going to be your best friend. They will think about you all day long when you are not home. You are the one that they care about more than anything and they put all of their trust in you. Give them the best chance possible at a happy and healthy life by always properly training them.

References

ASPCA. (n.d.). Pet Statistics. Retrieved from https://www.aspca.org/animal-homelessness/shelter-intake-and-surrender/pet-statistics

ASPCA. (n.d.) Vaccinations for Your Pet. Retrieved from https://www.aspca.org/pet-care/general-pet-care/vaccinations-your-pet

Becker, K. (2012). Want a Well-Behaved Dog? Do More of This and Less of That. Retrieved from https://healthypets.mercola.com/sites/healthypets/archive/2012/08/03/positive-reinforcement-dog-training.aspx

Bovsun, M. (2019). How to Potty Train a Puppy: A Comprehensive Guide for Success. Retrieved from https://www.akc.org/expert-advice/training/how-to-potty-train-a-puppy/

Dayton, R. (2016). New Study Shows Potential Benefits Of Spaying/Neutering Dogs After Age 1. Retrieved from https://pittsburgh.cbslocal.com/2016/11/14/new-study-shows-potential-benefits-of-spayingneutering-dogs-after-age-1/

Finlay, K. (2017). Can Dogs Taste? Retrieved from https://www.akc.org/expert-advice/lifestyle/can-dogs-taste/

Greenwood, A. (2016). Here Are The Surprising, Preventable Reasons Why People Give Up Their Pets. Retrieved from https://www.huffpost.com/entry/aspca-report-pets-given-up-to-shelters_n_56896846e4b06fa68882a134

Keliher, I. (n.d.). How to Choose the Perfect Dog Name (with Science).

Retrieved from **https://www.rover.com/blog/dog-name-advice/**

Manning, S. (2014). Behaviorists: Dogs feel no shame despite the look. Retrieved from https://www.usatoday.com/story/news/nation/2014/02/26/dogs-shame-guilty-look/5833395/

The Humane Society. (n.d.) Crate training 101. Retrieved from https://www.humanesociety.org/resources/crate-training-101

The Humane Society. (n.d.) Fact Sheet: Puppy Mills and Pet Stores. Retrieved from **https://www.humanesociety.org/sites/default/files/docs/pet-stores-puppy-mills-factsheet.pdf**

The Humane Society. (n.d.). Pets by the numbers. Retrieved from **https://www.animalsheltering.org/page/pets-by-the-numbers**

Walsh, K. (2017). Possible Health Isuses in Common Dog Breeds. Retrieved from https://www.healthline.com/health/dog-breeds-and-health-issues

Book 3 - Training Your Puppy Step-By-Step

www.ingramcontent.com/pod-product-compliance
Lightning Source LLC
Chambersburg PA
CBHW060822220526
45466CB00003B/942